AF703413

Extraterrestrial Prophecies & End-Time Warnings

Apocalyptic Texts on Cosmic Invasion or Salvation

A Modern Translation

Adapted for the Contemporary Reader

Various Ancient Writers

Translated by Tim Zengerink

© Copyright 2025
All rights reserved.

It is not legal to reproduce, duplicate, or transmit any part of this document in either electronic means or in printed format. Recording of this publication is strictly prohibited and any storage of this document is not allowed unless with written permission from the publisher except for the use of brief quotations in a book review.

This book contains works of fiction. Any resemblance to persons living or dead, or places, events, or locations is purely coincidental.

Table Of Contents

Preface - Message to the Reader

What If You Could Help Rebuild the Greatest Library in Human History?

Thousands of years ago, the Library of Alexandria stood as the crown jewel of human achievement — a sanctuary where the collected wisdom of every known civilization was gathered, preserved, and shared freely.

And then, it was lost.

Through fire, conquest, and the slow erosion of time, humanity lost not just books — but ideas, dreams, discoveries, and stories that could have changed the world forever.

Today, the Library of Alexandria lives again — and you are invited to be a part of its restoration.

Our mission is simple yet profound:

To rebuild the greatest library the world has ever known, and to translate all timeless works into every language and dialect, so that no seeker of knowledge is ever left behind again.

By joining our movement to rebuild the modern Library of Alexandria, you become part of an unprecedented mission:

- **Unlimited Access to the Greatest Audiobooks & eBooks Ever Written:**

  Instantly explore thousands of legendary works—Plato, Shakespeare, Jane Austen, Leo Tolstoy, and countless more. All instantly available to read or listen, placing a complete literary universe at your fingertips.

- **Beautiful Paperback & Deluxe Editions at Printing Cost**

  Own any title as an elegant paperback, deluxe hardcover, or stunning collectible boxset—offered to you at true printing cost, delivered straight to your door. Build your personal Library of Alexandria, crafted for beauty, built for durability, and worthy of proud display.

- **Fresh Translations for Modern Readers—in Every Language & Dialect**

  Enjoy timeless masterpieces reimagined in clear, contemporary language—no more outdated phrases or obscure references. Alongside the original versions, we're tirelessly translating these classics into every language and dialect imaginable, ensuring accessibility and understanding across cultures and generations.

- **Join a Global Renaissance of Literature & Knowledge**

  You directly support expanding our library, publishing deluxe editions at true cost, translating works into all global languages, and bringing humanity's greatest stories to people everywhere. By joining today, you're not just preserving a legacy of masterpieces; you set in motion a powerful wave of literary accessibility.

Become a Torchbearer of Knowledge.

Join us for free now at **LibraryofAlexandria.com**

Together, we will ensure that the light of human wisdom never fades again.

With gratitude and a shared love of knowledge,
The Modern Library of Alexandria Team

Visit:

www.libraryofalexandria.com

Or scan the code below:

Introduction

Unveiling the Cosmic Horizon: Ancient Texts and the Alien Imagination

For thousands of years, humans have gazed at the night sky and wondered: Are we alone? Are the heavens filled with divine beings, or are they populated with intelligences beyond our comprehension? For many religious traditions, the firmament is more than a canopy of stars—it is a realm teeming with angels, powers, and messengers. Yet as modern science ushered in the age of space travel and the search for extraterrestrial life, these spiritual beings began to be viewed through a new lens. Could the angels, watchers, and heavenly messengers of ancient scripture be descriptions of extraterrestrial entities? Could the cataclysmic events and divine interventions depicted in apocalyptic literature signal something more than symbolic prophecy? Could they be coded messages hinting at visitations from beyond the stars?

The texts collected in this volume—2 Baruch, 3 Baruch, 4 Ezra, the Apocalypse of Abraham, the Vision of Ezra, and selections from the Sibylline Oracles—form a tapestry of ancient eschatological speculation. They were written in the crucible of despair and hope, often in response to real historical calamities such as the destruction of the Jerusalem Temple. Their authors looked forward to a future restoration, but that restoration was often mediated by mysterious beings who descended from the heavens, wielding divine judgment and cosmic power. While mainstream religious interpretation has traditionally understood these texts in spiritual or allegorical terms, an increasing number of researchers, theologians, and thinkers have begun to ask what happens when we take their celestial imagery at face value. What if these ancient visionaries were not just dreaming of metaphysical truths—but recording encounters with otherworldly visitors?

Interstellar Apocalypse: Reading End-Time Texts Through a New Lens

The idea that extraterrestrials might play a role in humanity's destiny is not new. From Erich von Däniken's seminal 1968 work Chariots of the Gods? to the speculative philosophies of modern UFO religions, the notion that ancient gods were actually spacefaring beings has taken deep root in the public imagination. Within biblical and apocryphal traditions, apocalyptic literature has always served as a crucible for cosmic speculation. These texts are obsessed with endings—of history, of empires, of time itself—and with transitions to higher orders of existence. Often encoded in their vivid language are themes of great upheaval, aerial intervention, fiery purgation, and messianic deliverance. In other words, they already speak the symbolic language of cosmic change.

When 2 Baruch speaks of a coming judgment and of those who "descend from above," is it possible that its author tapped into a deeper collective memory of sky-borne beings—those who may have once visited, or who might yet return? When 4 Ezra imagines a mysterious messianic figure rising from the sea and commanding the nations, does it merely symbolize divine intervention, or does it invoke something more akin to an alien arrival—an emissary of a higher order of intelligence appearing at humanity's hour of need? These are the kinds of questions that animate both the ancient apocalyptists and their modern interpreters.

The Apocalypse of Abraham is perhaps the most striking in this context. It describes a visionary journey through multiple heavens, filled with strange beings and cryptic revelations. Its portrayal of Azazel, a fallen spirit associated with corruption, resonates strongly with the figure of the Watchers in 1 Enoch—beings who descend to Earth, teach forbidden knowledge, and breed a race of giants. The Vision of Ezra similarly presents a sweeping cosmic drama where the boundaries between earth and heaven are porous, and divine agents wield judgment with supernatural force. The Sibylline Oracles, too, offer dire warnings wrapped in cosmic language, as celestial signs and heavenly fire signal the world's impending transformation.

In reading these works today, one can see a dual trajectory. On the one hand, they express timeless human anxieties: fear of corruption, longing for justice, and the hope for renewal. On the other hand, they provide an interpretive canvas onto which we can project modern concerns—particularly the possibility that our species is not alone in the cosmos, and that our fate might someday be entwined with forces that descend from the stars.

Between Myth and Revelation: Ancient Prophecies for a New Age

The question remains: were these authors merely speaking in metaphor, or were they preserving genuine encounters—interpreted in the only language available to them? It is easy to dismiss apocalyptic literature as a product of myth, fear, or psychological trauma. And yet, many modern readers find that these texts ring with a strange familiarity. Their descriptions of beings who arrive in clouds, of fiery vehicles, of transformations and resurrections, feel oddly congruent with 21st-century concepts of interdimensional travel, advanced technology, and even alien abduction narratives.

This is not to suggest that the ancient prophets and scribes were writing science fiction. Rather, it is to acknowledge that their language may have encoded real encounters—cosmic or spiritual—in the mythic idioms of their time. The overlap between mystical experience and contact phenomena has been widely noted in both religious and ufological circles. Many so-called "experiencers" today describe journeys through light, contact with radiant beings, downloads of knowledge, and transformative visions not unlike those reported in these ancient texts.

Furthermore, the ethical and theological concerns embedded in these works remain profoundly relevant. Whether the messianic figure of 4 Ezra is a heavenly warrior, a divine judge, or a cosmic traveler, the call to righteousness and repentance echoes through the ages. Whether Azazel represents spiritual corruption or alien tampering with human evolution, the cautionary tale persists: humanity must discern its place within a larger cosmic order, whether terrestrial or extraterrestrial.

In bringing together these ancient texts under the lens of extraterrestrial speculation, this volume does not seek to impose a singular interpretation. Rather, it opens the door to a new hermeneutic—one that embraces the mystery, the imagination, and the possibility that the universe is not only more vast than we imagine, but more intimately connected to our spiritual heritage than we dare believe. Let the reader decide: were the prophets witnessing angels or aliens? Was divine fire symbolic, or was it the afterglow of technologies lost to time? Were the heavens merely the dwelling place of God—or the cosmic frontier where destiny, judgment, and redemption await?

The apocalypse, after all, means unveiling. And in this unveiling, ancient voices may yet reveal the shape of things to come.

The Second Book of Esdras

This is the second book of the prophet Esdras, a descendant of Aaron from the tribe of Levi. Esdras lived as a captive in the land of the Medes during the reign of King Artaxerxes of Persia.

Chapter 1

The Lord spoke to me, saying, "Go and tell my people about their sins. Show their children the wrongs they have committed against me so they can warn future generations not to repeat them. Their sins are even greater than those of their ancestors because they have forgotten me and worshiped foreign gods. Didn't I free them from slavery in Egypt? Yet, they have turned away from me and made me angry.

Mourn for them and bear the weight of their sins, for they have rejected my laws and become a rebellious nation. After all I have done for them, how much longer should I have patience with them? I brought down powerful kings for their sake. I struck down Pharaoh and his entire army. I defeated many nations before them, scattered the people of Tyre and Sidon, and conquered all their enemies.

Now tell them, 'The Lord says: I made a path through the sea for you. I led you where there was no road. I chose Moses to guide you and Aaron to serve as your priest. I gave you light through a pillar of fire and performed great miracles, yet you still turned your back on me,' says the Lord.

'I, the Almighty, sent quails for you to eat and provided a safe place for your camp, yet you still complained. You didn't thank me when I defeated your enemies, and even now, you continue to grumble. Where are the blessings I poured out on you? When you were hungry and thirsty in the wilderness, didn't you cry out, saying, "Why did you bring us here to die? We would have been better off as slaves in Egypt." But I had mercy on you and gave you manna, the bread of angels.

When you were thirsty, didn't I split open a rock to give you water? When the sun's heat was too much, didn't I shade you with tree leaves? I gave you fertile land and drove out the Canaanites, Perizzites, and Philistines. What more could I have done for you?' says the Lord.

'I, the Almighty, did not punish you when you spoke against me at the bitter waters. Instead, I made the water sweet by placing a tree in it. What am I to do with you, Jacob? And Judah, why won't you listen? I will turn to other nations and give them my name so they will obey my laws. Since you have rejected me, I will reject you. When you cry out for help, I will not answer.

When you beg for mercy, I will not listen because your hands are covered in blood, and you rush to do wrong. You have not only abandoned me—you have destroyed yourselves,' says the Lord.

'I have called you like a father calls his sons, like a mother calls her daughters, and like a caregiver watches over children. I wanted you to be my people so I could be your God. I wanted you to be my children so I could be your father. I gathered you like a hen gathers her chicks under her wings. But now, what else can I do? I will remove you from my presence. When you offer sacrifices, I will turn away. I have rejected your feasts, your celebrations, and even your traditions.

I sent prophets to guide you, but you killed them and ignored their message. I will hold you accountable for their blood,' says the Lord. 'Your land will be left empty. I will scatter you like dust in the wind. Your children will not prosper because they rejected my commandments and chose to do evil.

I will give your homes to a people who have not yet come, a people who do not know me now but will believe in me. They have not seen the miracles you have seen, yet they will obey my laws. Even without prophets, they will remember their sins and turn back to me. I call upon these future people, whose descendants will rejoice and give thanks. Even though they will not see me with their eyes, they will believe in my words through the Spirit.'

Now, father, look and take joy as you see the people coming from the east. I will give them leaders: Abraham, Isaac, and Jacob, along with Hosea, Amos, and Micah, Joel, Obadiah, and Jonah, Nahum, and Habakkuk, Zephaniah, Haggai, Zechariah, and Malachi, who is also called the Lord's messenger."

Chapter 2

The Lord says, "I rescued these people from slavery and gave them my laws through my prophets, but they refused to listen and ignored my commands. Their mother mourns, saying, 'Go on your way, my children, for I am left alone, abandoned like a widow. I raised you with joy, but now I have lost you in sorrow and pain because you sinned against the Lord and acted wickedly before me.

What else can I do for you now? I am left desolate, abandoned. Go, my children, and ask the Lord for mercy.' But as for me, O Father, I call on you as a witness, along with their mother, because these children have broken my covenant. May they face shame, and may their mother be ruined, leaving no descendants behind. Let them be scattered among the nations, and let their names be erased from the earth, for they have rejected my covenant.

Woe to you, Assur, who shelter the wicked! You sinful nation, remember what I did to Sodom and Gomorrah. Their land was turned into burning pitch and ashes. I will do the same to those who refuse to obey me," says the Lord Almighty.

The Lord told Esdras, "Tell my people that I will give them the kingdom of Jerusalem, the kingdom I had once prepared for Israel. I will remove their shame and give them eternal homes that I have made for them. They will enjoy the tree of life and its sweet fragrance. They will no longer struggle or grow weary.

Ask, and you will receive. Pray that your days may be shortened, for the kingdom is already prepared for you. Stay alert! Call upon heaven and earth as witnesses. Let them testify that I have removed evil and created what is good, for I am the living God, says the Lord.

Mother, embrace your children. I will bring them to you with joy, like a dove carrying her young. Strengthen their steps, for I have chosen you, says the Lord. I will raise the dead from their graves and bring them back to life because they bear my name. Do not be afraid, mother of children, for I have chosen you, says the Lord.

To guide you, I will send my servants Isaiah and Jeremiah. Through them, I have prepared twelve fruitful trees for you, along with streams flowing with milk and honey, and seven great mountains where roses and lilies grow in abundance. I will fill your children with joy in this place.

Defend widows. Protect orphans. Help the poor. Stand up for those who are helpless. Clothe those in need. Heal the brokenhearted and the weak. Do not mock those with disabilities. Help the injured. Let the blind see my glory. Care for the elderly and the young among you. When you find the dead, mark their graves and lay them to rest, and I will honor you in my resurrection.

Rest now, my people, and find peace, for your time of rest is near. Care for your children, faithful caretaker, and strengthen them. Of the servants I have given you, not one will be lost, for I will call them back to me. Do not be afraid, for in the days of suffering, while others weep and mourn, you will rejoice and be filled.

The nations will envy you, but they will not harm you, says the Lord. My hands will protect you so that your children will not see destruction. Rejoice, mother, with your children, for I will rescue you, says the Lord. Remember your children who rest, for I will bring them back from their hidden places in the earth and show them mercy, for I am a merciful God, says the Lord Almighty. Hold your children close until I come, and speak to them of my mercy, for my blessings overflow, and my kindness never ends."

I, Esdras, was commanded by the Lord on Mount Horeb to go to Israel. But when I went, they rejected both me and the Lord's commandments. So now I say to you, O nations who hear and understand: Seek your shepherd. He will give you eternal rest, for the one who is coming at the end of time is near. Be ready to receive the rewards of the kingdom, for eternal light will shine upon you forever.

Turn away from the darkness of this world and embrace the joy of your glory. I openly call on my Savior as a witness. Accept the gifts the Lord has given you, and rejoice, giving thanks to Him who has called you into His heavenly kingdom. Rise and stand, and look at the great crowd of those who have been marked for the Lord's feast. These are the ones who left behind the ways of this world and have been dressed in glorious robes by the Lord.

Zion, take back your full number and complete the count of those clothed in white, those who have followed the law of the Lord. The children you have longed for are now complete in number. Pray to the Lord Almighty to set apart the people He has chosen from the beginning."

I, Esdras, saw on Mount Zion a vast crowd, too great to count, all praising the Lord with songs. In the middle of them stood a young man, taller than all the rest, placing crowns on their heads. He was more honored than anyone else, and I was amazed by him.

I asked the angel, "Who are these people, my lord?" The angel replied, "These are the ones who set aside their earthly lives and took on immortality. They declared the name of God, and now they are crowned and receive palms as their reward."

Then I asked, "Who is the young man placing crowns on them and giving them palms in their hands?" The angel answered, "He is the Son of God, the one they confessed while they lived in the world."

At that moment, I began to praise those who had stood strong for the name of the Lord. Then the angel said to me, "Go and tell my people about the great and wonderful things of the Lord God that you have seen."

Chapter 3

Thirty years after the city's destruction, I, Salathiel, also known as Esdras, was in Babylon. I felt troubled and overwhelmed with thoughts. As I lay on my bed, I reflected on the ruins of Zion and the success of those living in Babylon. My heart was heavy, and in fear, I turned to the Most High and prayed:

"O Lord, you spoke the world into existence from the very beginning. You shaped the earth and commanded the dust to take form. From that dust, you created Adam, molding him with your own hands. You breathed life into him, making him a living being. You placed him in the garden you planted before the earth was fully formed. You gave him only one simple command, yet he disobeyed. Because of this, you decreed death for him and his descendants.

From Adam came many nations, tribes, and countless people. Each nation followed its own ways, choosing sin over your commandments. Yet, you allowed them to continue. Eventually, you sent a great flood to wipe humanity from the earth. Just as Adam had died, so did they. But you spared Noah and his family, and from them came the righteous who survived.

When the earth was repopulated, people increased in number again, forming families, nations, and tribes. But their hearts became even more corrupt than before. When they acted wickedly, you chose one man, Abraham, from among them. You loved him and revealed to him the mysteries of the future. You made an everlasting promise to him, swearing never to abandon his descendants. You gave him Isaac, and from Isaac, you gave Jacob and Esau. You set Jacob apart for yourself but rejected Esau. From Jacob came a great nation.

You led Jacob's descendants out of Egypt and brought them to Mount Sinai. You opened the heavens, shook the earth, and caused the deep waters to tremble. Your glory passed through four great forces—fire, earthquake, wind, and ice—so that you could give your law to Jacob's children and commandments to the people of Israel.

Yet, you did not remove their sinful hearts, so your law could not take root within them. The first man, Adam, had a corrupt heart and disobeyed, and so did all his descendants. Sin became deeply rooted. Even though you gave them your law, their sinful nature remained. Goodness faded, and evil increased.

As time passed and the end grew closer, you raised up your servant David. You commanded him to build a city where your name would be honored and where sacrifices would be made from the gifts you had given. For many years, your people followed this practice. But in time, the people of the city became corrupt, just as Adam and his generations had before them. Because their hearts were sinful, they turned to wickedness. So you allowed their enemies to overtake the city.

I asked myself, 'Are the people in Babylon better than those in Zion? Is that why they now rule over Zion?' But when I arrived here, I saw even greater evil. Now, in this thirtieth year, my heart is broken because I see so much sin. I see how you allow these sinners to continue their wickedness while you destroy your own people. You protect those who do not follow you and punish those who do.

You have not shown me how your justice works. Are the actions of Babylon better than those of Zion? Is there any nation besides Israel that truly knows you? Is there any other tribe as faithful to your covenant as the descendants of Jacob? Yet their devotion seems to go unrewarded, and all their efforts seem meaningless. I have traveled among other nations and seen their wealth, but they do not care about your commandments.

Now, weigh our sins against those of the rest of the world. See where the balance falls. Has anyone on earth not sinned before you? What nation has fully obeyed your laws? Perhaps a few individuals have remained faithful, but an entire nation? There is none."

Chapter 4

The angel sent to me, whose name was Uriel, said, "Your understanding of this world is limited. Do you really think you can comprehend the ways of the Most High?"

I replied, "Yes, my lord."

He said, "I have been sent to show you three things and ask you three questions. If you can answer even one of them, I will show you the path you seek and explain why the human heart leans toward evil."

I said, "Please, ask your questions, my lord."

He said, "Weigh the weight of fire, measure the wind, or recall a single day that has already passed."

I answered, "No human could ever do that. Why ask me such impossible things?"

He said, "If I had asked you how many homes lie beneath the sea, how many springs feed the ocean, how many rivers flow above the sky, where the gates of the underworld are, or the paths to paradise, you might have said, 'I have never gone into the abyss, nor have I seen the underworld, nor have I ascended to heaven.' But I only asked you about things you encounter daily—fire, wind, and time—yet even those, you cannot explain."

He continued, "If you cannot understand the things around you, things you see and experience every day, how can you hope to grasp the ways of the Most High? How can someone living in a world full of corruption comprehend things that are beyond corruption?"

When I heard this, I fell to the ground and said, "It would have been better if we had never been born than to live in a broken world, suffer, and remain ignorant of our purpose."

He answered, "Let me give you an example. The trees of the forest once gathered and made a plan. They said, 'Let's rise up and fight against the sea so it will move back and we can expand our land.' Likewise, the waves of the sea also made a plan and said, 'Let's rise up and conquer the land so we can expand our waters.' But neither succeeded—the trees were burned by fire, and the waves were held back by the sand.

"If you were their judge, which would you say was right, and which was wrong?"

I answered, "Both were foolish. The land was made for the trees, and the sea was made for the waves."

Then he said, "You have judged wisely. So why do you not apply the same understanding to your own situation? Just as the earth belongs to the trees and the sea belongs to the waves, humanity was meant to understand earthly things, while the One who dwells in heaven understands heavenly matters."

I replied, "My lord, then why was I given the ability to reason? I am not asking about things beyond me, but about what happens around us every day. Israel has been handed over to godless nations. The people you loved have been given to those who do not know you. The law of our ancestors is no longer followed, and your promises seem forgotten. We disappear like shadows. Our lives are short and meaningless, and we are unworthy of mercy. What will you do for your name, the name by which we are called? This is what I ask."

He replied, "If you live, you will see it. If you endure, you will be amazed, for this world is quickly passing away. It cannot hold the things promised to the righteous in the future because it is full of sorrow and weakness. The evil you ask about has been planted, but the time to harvest it has not yet come. If the wickedness that has grown is not removed first, and if this world does not pass away, then the new world where goodness will flourish cannot appear.

"From the beginning, a single seed of evil was planted in Adam's heart, and look how much corruption has come from it even to this day! Imagine how much worse it will grow until the time of harvest. If one seed of wickedness has produced so much, how great will the final harvest of evil be when so many seeds have been sown?"

I asked, "How long must we wait? When will these things happen? Why is life so short and full of suffering?"

He responded, "Do not try to rush the plans of the Most High. Your impatience is for yourself, but God works for the good of many. Even the souls of the righteous have asked the same question while waiting in their chambers, saying, 'How long must we wait? When will the time of the harvest come?' To them, Jeremiel the archangel answered, 'You must wait until the full number of righteous souls is complete. The Most High has carefully measured time, counted the seasons, and set everything in place. Nothing will change or move until the appointed time is fulfilled.'"

I said, "O Lord, we are all full of sin. Could it be that the delay in the gathering of the righteous is because of the sins of those who still live on earth?"

He replied, "Go and ask a pregnant woman if she can hold her child in the womb beyond nine months."

I answered, "No, Lord, she cannot."

He said, "In the same way, the chambers holding the souls in the underworld cannot keep them beyond the appointed time. Just as a woman in labor delivers her child when the time comes, so too will these places release what they hold when the time is fulfilled. Then you will see the things you long to understand."

I said, "If I have found favor in your sight, please tell me this: Is the time that has already passed greater, or is the time that remains ahead greater? I can see what has happened, but I do not know what is still to come."

He said, "Stand on my right side, and I will show you the answer."

So I stood, and I saw a blazing fire pass before me. After the flames disappeared, only smoke remained. Then a dark cloud appeared, heavy with rain, pouring down a great flood with a storm. When the storm ended, only a few drops of water remained.

Then he said to me, "Think about this: Just as the rain is far greater than the few drops left behind, and just as the fire is much stronger than the smoke that remains, so too, the time that has already passed is far greater than what is still to come."

I prayed and asked, "Will I live to see that time? Or who will be alive to witness those days?"

He replied, "I can tell you some of the signs you seek, but I was not sent to reveal how long you will live, for I do not know."

Chapter 5

"But understand this about the signs: The time will come when the people on earth will be filled with fear and confusion. Truth will be hidden, and faith will disappear. Evil will spread far beyond anything you have ever known. The land that now seems strong and successful will become a barren wasteland, empty for all to see.

But if the Most High allows you to live, you will witness what happens after the third period of time. The sun will suddenly shine at night, and the moon will glow during the day. Trees will drip with blood, and stones will cry out. People will be overcome with terror, and stars will fall from the sky.

A leader will rise whom no one expected, and birds will flee together in large numbers. The sea near Sodom will throw out fish and make strange, loud noises at night—sounds no one has heard before, but everyone will hear. Chaos will spread everywhere. Fires will break out more often, wild animals will wander into strange places, and women will give birth to unnatural creatures.

Saltwater will mix with fresh, and even close friends will betray one another. Wisdom will disappear, and understanding will hide itself. Many will search for it desperately, but they will not find it. Wickedness and a lack of self-control will take over the earth.

One nation will ask another, 'Has righteousness passed through here? Has anyone who does good come this way?' And the answer will be, 'No.' In those days, people will hope for a better future, but their hopes will be empty. They will work hard, but their efforts will bring no reward.

I am allowed to tell you these signs. If you pray, cry out to God, and fast for seven more days, you will receive even greater revelations."

When I woke up, my body trembled violently, and my mind was so distressed that I felt weak. The angel who had spoken to me came, supported me, comforted me, and helped me stand.

On the second night, Phaltiel, a leader among the people, came to me and asked, "Where have you been? Why do you look so troubled? Don't you realize that Israel depends on you, even in this land of captivity? Get up, eat something, and don't abandon us—like a shepherd who leaves his flock to be eaten by wolves."

I answered, "Leave me alone. Do not come near me for seven days. After that, you may return." He respected my request and left.

I fasted for seven days, mourning and weeping, just as the angel Uriel had told me to do. After the seven days, my heart was heavy again, but my spirit was renewed with understanding. I spoke once more to the Most High:

"O Lord, out of all the forests in the world, you chose one special vine. From all the lands on the earth, you picked one particular country. Out of all the flowers that exist, you selected one lily.

From all the waters in the seas, you set one river apart for yourself. Out of all the cities ever built, you made Zion your holy place. From all the birds in the sky, you chose one dove as your own. Among all the animals on earth, you picked one sheep.

And from all the nations on earth, you chose one people to be yours. You gave this special people, whom you love, a law that is approved by all. But now, Lord, why have you handed them over to so many others? Why has this precious people been dishonored and scattered among countless nations?

Those who reject your promises have trampled on those who trust in your covenant. If you are so angry with your people, why not punish them with your own hand instead?"

As I finished speaking, the angel who had visited me before returned and said, "Listen carefully, and I will teach you. Pay close attention, and I will explain further."

I said, "Speak, my lord."

Then he said, "You are deeply troubled about Israel. But do you love this people more than the One who created them?"

I answered, "No, my lord, but I am deeply troubled for my people. My heart aches as I try to understand the ways of the Most High and search for even a small glimpse of His judgment."

He said to me, "You cannot understand it."

I asked, "Why, Lord? Why was I even born? Why couldn't my mother's womb have been my grave, so I wouldn't have to see the suffering of Jacob and the hardships of Israel's people?"

He replied, "If you can count those who have not yet been born, gather all the scattered drops of water from the sea, make dried flowers bloom again, or open sealed places and release the winds trapped inside, then I will explain the troubles you wish to understand."

I said, "O Lord, only someone who does not live among mortals could know such things. As for me, I have no wisdom. How could I possibly understand the things you ask?"

He said, "Just as you are unable to do the things I have asked, you cannot comprehend my judgment or the depth of the love I have for my people."

I asked, "But Lord, you have made promises to those who will live at the end of time. What about those who came before us, those alive now, and those yet to come?"

He answered, "Think of my judgment like a ring. Just as those who come last do not arrive late, neither are those who come first made to arrive early."

I said, "Couldn't you have created all people—those who have lived, those alive now, and those who will live in the future—all at once, so your judgment could come sooner?"

He replied, "Creation cannot move ahead of its Creator, and the world cannot hold everyone at once who has ever lived or will live."

I said, "But you said that one day all creation will come to life together. If the world will be able to sustain them all at once in the future, why can't it do so now?"

He answered, "Ask a woman's womb this question. Say to her, 'If you carry ten children, why don't you give birth to them all at the same time?' She will tell you that it is impossible, for each child must be born in its own time."

I replied, "Of course, she cannot. Each child must come when the time is right."

Then he said, "In the same way, I have given the earth a womb. Those who are meant to be born will come forth at their appointed time. Just as a woman cannot bear children after she has grown old, so too I have arranged the world to follow its course."

I asked, "Since you have explained this to me, may I ask one more question? The mother you spoke of—is she still young, or is she reaching the end of her strength?"

He answered, "Ask a woman who has had children, and she will explain it to you. Say to her, 'Why aren't the children you bear now as strong and large as those you had when you were younger?' She will tell you, 'The children I gave birth to in my youth were different from those born when my body became older and weaker.'

Think about this yourself. You are smaller and weaker than those who came before you, and those who come after you will be even smaller, for they are born into a world that is aging and has passed its prime."

I said, "Lord, I beg you, if I have found favor in your sight, please reveal to me who you have chosen to care for your creation."

Chapter 6

He said to me, "Before the earth was created, before the gates of the world were set in place, before the winds began to blow, before thunder and lightning existed, before paradise was formed, before flowers bloomed or earthquakes shook the land, before the great army of angels was gathered, before the sky stretched above, before the dimensions of the heavens were measured, and before the foundation of Zion was laid—before time itself began, before those who sin existed, and before those who store up faith were chosen—at that moment, I had already thought of these things. I created everything by myself, and just as I brought them into being, I will bring them to an end."

I asked, "How will we know when this age is ending and the new one is beginning?"

He answered, "Think of Jacob and Esau. When Jacob was born, he grabbed Esau's heel. Esau represents the end of this age, and Jacob represents the beginning of the next. A person's birth begins with the hand, and their life ends at the heel. So do not search beyond the hand and the heel, Esdras."

I said, "Lord, if I have found favor with you, please show me how the signs you have revealed to me will come to an end."

He replied, "Stand up, and you will hear a powerful voice. If the ground beneath you shakes when it speaks, do not be afraid, because that voice is announcing the end of all things. The foundations of the earth will recognize it and tremble, for they know their time of change has come."

I stood up and listened. Suddenly, I heard a voice like rushing water. It said, "The time is coming when I will visit those who live on the earth. I will examine those who have caused harm through their injustice. When Zion's suffering is complete and when this age has run its course, I will show these signs: The heavens will open, and books will be displayed for all to see.

Children who are only a year old will suddenly speak with clear voices. Pregnant women will give birth at three or four months, and their babies will live and grow. Farmlands that were once full will seem as though nothing was ever planted there. Storehouses filled with food will suddenly be empty.

A loud trumpet will sound, and when people hear it, they will be overcome with fear. Friends will turn against each other as if they were enemies. The earth will shake along with its people. The springs of water will stop flowing, and for three hours, no water will be found.

Those who survive all these things will be saved. They will witness my salvation and the end of this world. They will see those who were taken up—those who never experienced death from the moment they were born. The hearts of those who remain will be transformed, and they will be given a new spirit.

Evil will be destroyed, and lies will disappear. Faith will grow strong, corruption will be defeated, and truth, which has been hidden for so long, will finally be revealed."

As the voice spoke, I felt the ground beneath me slowly shifting. The angel said, "I have come tonight to show you these things. If you pray and fast for seven more days, I will reveal even greater things than what I have shown you. Your voice has reached the Most High. The Mighty One has seen your righteousness and purity, which you have kept since your youth. That is why He sent me to show you these things and to tell you, 'Believe and do not be afraid! Do not focus too much on the past, or you will rush too quickly toward the future.'"

After hearing this, I wept again and fasted for seven more days, just as I had been instructed. On the eighth night, my heart became troubled again, and I prayed to the Most High. My spirit was stirred, and my soul was heavy with sorrow. I said,

"O Lord, at the beginning of creation, on the first day, you spoke, saying, 'Let heaven and earth be made,' and by your command, they were formed. Your Spirit hovered, and everything was covered in darkness and silence. No human voice could yet be heard. Then you commanded light to shine, so your works could be seen.

On the second day, you created the sky and separated the waters—some above and some below. On the third day, you commanded the waters to gather into one part of the earth, leaving the rest dry so it could be planted and cultivated to serve you. As soon as you spoke, it was done. Instantly, plants, fruits of every kind, flowers of unmatched beauty, and sweet fragrances appeared.

On the fourth day, you ordered the sun to shine, the moon to glow, and the stars to follow their courses. You assigned them to serve the humans who were to be created. On the fifth day, you commanded the waters to produce living creatures—fish, birds, and all sea creatures. The still and lifeless waters suddenly brought forth life, so that all nations might marvel at your wonders.

You also set apart two great creatures. One you named Behemoth, and the other you called Leviathan. You separated them because the waters of the earth could not hold them both. You gave Behemoth the land created on the third day, a place of many hills. Leviathan was placed in the deep waters. You have kept them ready for when you decide to use them.

On the sixth day, you commanded the earth to produce livestock, wild animals, and all creatures that crawl. Then you created Adam and made him ruler over all your works. From him came all of us—the people you have chosen.

Lord, I say all this before you because you declared that you created this world for our sake. As for the other nations that also came from Adam, you have said they are like nothing, like a drop falling from a bucket, like a speck of dust.

Now, Lord, look at these nations, which are counted as nothing, yet they rule over us and destroy us. But we, your people—your firstborn, your beloved children—have been handed over to them. If the world was made for us, why do we not possess it as our inheritance? How much longer must we wait?"

Chapter 7

When I finished speaking, the angel who had visited me before came to me again. He said, "Stand up, Esdras, and listen to the words I have been sent to share with you."

I replied, "Speak, my lord."

Then he said, "Imagine a vast sea, open and endless, but the only way to reach it is through a narrow river. If someone wants to enter the sea to explore it or take control of it, they must first pass through the narrow entrance. If they refuse to go through the difficult path, how could they ever reach the wide expanse?

Now think of a city built on a flat plain, filled with all kinds of good things, but the only way in is through a narrow and dangerous passage. On one side, there is fire, and on the other, deep water. The path between them is so tight that only one person can pass through at a time. If someone is meant to inherit this city, how could they claim it unless they first pass through the dangers?"

I said, "That makes sense, Lord."

Then he said to me, "This is exactly how Israel's inheritance works. I created the world for their sake, but when Adam disobeyed my command, the way to this world became filled with hardships, sorrow, and danger. The entrances are few and difficult. But the way to the greater world is wide, safe, and leads to eternal life.

If people do not go through the struggles of this life, they will never reach the rewards I have prepared for them. So why are you troubled, knowing that you are only human? Why are you so shaken, considering that your time here is temporary? Why haven't you focused more on the future instead of worrying only about the present?"

I answered, "O Lord, in your law you said that the righteous will receive these blessings, but the wicked will perish. The righteous suffer now with the hope of peace later, while the ungodly suffer in this life and will never find relief."

He replied, "You cannot judge better than God, and you do not have greater wisdom than the Most High.

It is better for many who live now to perish than for God's law to be ignored. He gave clear instructions to those who entered the world about how to live, how to gain life, and how to avoid punishment. But they refused to listen. They spoke against Him, followed their own empty ideas, and created plans for evil. Some even claimed that the Most High doesn't exist. They ignored His ways, rejected His laws, denied His promises, refused His commands, and did nothing He required of them.

So, Esdras, those who have chosen emptiness will receive nothing, while those who have followed the truth will receive what is full. The time is coming when the signs I told you about will happen. Then the bride—the holy city—will appear. She will be revealed to everyone, even though she is now hidden from the world. Those who survive the troubles I have warned you about will see my wonders."

"For my son, Jesus, will be revealed, along with those who are with him, and they will rejoice for four hundred years. After that, my son, the Anointed One, will die, and so will all living people. Then the world will return to complete silence for seven days, just as it was in the beginning, and no one will remain alive.

After those seven days, the world will awaken again, and everything corrupt will be destroyed. The earth will release the dead buried within it, the dust will give up those who have slept in silence, and hidden places will return the souls that were entrusted to them. The Most High will appear, seated for judgment. Mercy will no longer be given, and patience will end.

Only judgment will remain. Truth will stand strong. Faith will be firm. Rewards will be given. Good deeds will be remembered, and wickedness will no longer be hidden. The pit of punishment will be opened, and next to it, the place of rest. The fire of hell will be revealed, and near it, the paradise of joy."

Then the Most High will say to the nations that have been raised from the dead, "Look and see the One you rejected, the One you refused to serve, and the commandments you chose to ignore. Look to your left and right—on one side is joy and rest, and on the other is fire and torment." These are the words He will speak on the day of judgment.

On that day, there will be no sun, no moon, no stars, no clouds, no thunder, no lightning, no wind, no water, no air. There will be no darkness, no morning or night, no summer or spring, no heat or winter. No frost, no cold, no hail, no rain, no dew. There will be no noon, no dawn, no brightness or light—only the overwhelming glory of the Most High. Through this light, everyone will clearly see what is before them.

This will last as though it were a week of years. This is my judgment, but I have only shown these things to you."

I said, "O Lord, I have said it before, and I will say it again: Blessed are those who are alive and follow your commandments! But what about those I have prayed for? Who among the living has not sinned? Who among humanity has not broken your covenant? Now I see that the world to come will bring joy to only a few, but suffering to many.

An evil heart has grown inside us, pulling us away from your commandments. It has corrupted us and led us toward death. It has pointed us toward destruction and taken us far from life. This has not happened to just a few—it has happened to almost everyone who has ever lived."

He replied, "Listen to me, and I will teach you. Let me correct you again. The Most High did not create just one world, but two.

You said that the righteous are few and the wicked are many. Here is why: If you had a handful of rare jewels, would you mix them with worthless lead and clay?"

I answered, "No, Lord, that would not make sense."

He said, "Exactly. Now ask the earth, and it will teach you. Let it guide you, and it will help you understand. Tell the earth, 'You produce gold, silver, bronze, iron, lead, and clay. But silver is more common than gold, bronze is more abundant than silver, iron is more plentiful than bronze, lead is more common than iron, and clay is the most common of all.' Now, judge for yourself—what is more valuable, things that are rare or things that are plentiful?"

I responded, "Lord, things that are common have little worth, but rare things are treasured far more."

He answered, "Think deeply about this. When someone has something that is rare and difficult to obtain, they value it much more than what is easy to find. This is how my judgment works. I take joy in the few who are saved because they glorify my name and bring honor to it. Through them, my name is respected. But I do not grieve for the multitudes who perish, for they are like mist that vanishes or flames that burn brightly for a moment before fading into smoke."

I said, "O earth, why do you bring forth people, when the human mind—made from the same dust as everything else—causes so much suffering? Wouldn't it have been better if the dust had never been shaped into a human mind, so we would not have to endure this pain? As the mind grows, it burdens us with the knowledge of our own mortality, and this awareness brings us sorrow.

Let humanity mourn, but let the animals of the earth rejoice. Let those who are born cry, but let the beasts of the wild be glad. Their lives are better than ours because they face no judgment. They do not suffer torment, nor do they long for salvation. What is the purpose of life if it only leads to suffering?

Everyone who is born carries the weight of sin, burdened by guilt and wrongdoing. If judgment after death did not exist, perhaps we would be better off."

He replied, "When the Most High created the world, Adam, and all who came from him, He also prepared the judgment that would follow.

You are right that the human mind grows with understanding. Those who live on earth experience pain because, despite having knowledge, they still choose to sin. They were given commandments, yet they did not obey them. The law was given to them, but they were not faithful to it. So what can they say when judgment comes? How will they defend themselves in the last days?

The Most High has shown patience toward those who live on earth—not because of their actions, but because of the times He has determined."

I asked, "Lord, if I have found favor with you, please tell me this: When a person dies and their soul leaves their body, will they find rest until the time when you renew creation, or will they face torment right away?"

He answered, "I will explain this to you. But do not align yourself with those who reject God, nor count yourself among those who are destined for punishment. You have stored up a treasure of good deeds with the Most High, though it will not be revealed to you until the end times.

As for death, here is what happens: When the Most High decides that a person must die, their spirit leaves their body and returns to the One who gave it. First, the spirit praises the glory of the Most High. But if that person rejected God, despised His law, and hated those who feared Him, their spirit will not find rest. Instead, it will wander in grief and suffering, facing seven different kinds of torment."

The first torment is knowing they rejected the Most High's law. The second is realizing they lost their chance to repent and gain eternal life. The third is seeing the rewards prepared for those who followed God's ways. The fourth is understanding the suffering that awaits them at the end. The fifth is witnessing the peaceful homes of the righteous, protected by angels. The sixth is knowing their punishment is coming soon. The seventh, and worst of all, is the overwhelming shame, confusion, and fear they will feel as they see the glory of the Most High—the One they rejected and must now face in judgment.

This is what will happen to those who lived in obedience to God when their souls leave their bodies. During their lives, they remained faithful, enduring hardships to follow His law. For them, there will be joy in seven ways.

First, they will rejoice because they resisted evil and did not let it destroy them. Second, they will be glad as they witness the downfall of the wicked. Third, they will be at peace knowing their Creator has seen their faithfulness. Fourth, they will rest safely, surrounded by angels, while looking forward to the glory that awaits them. Fifth, they will celebrate because they have escaped corruption and gained the promise of eternal life. They will see the suffering they avoided and rejoice in their freedom. Sixth, their faces will shine like the sun, knowing they will never grow old or weak again. Seventh, and most important of all, they will stand before the Most High with confidence, without fear or shame, and receive their eternal reward in His presence.

This is the destiny of the righteous. These are the torments of those who rejected the Most High. I have told you these things.

I asked, "After someone dies, will they experience all of this immediately?"

He replied, "They will have seven days to see and understand these things before they are taken to their final place."

I said, "Lord, if I have found favor with You, please answer this: On the Day of Judgment, will the righteous be able to pray for the wicked? Will parents intercede for their children, children for their parents, relatives for each other, or friends for those they love?"

He replied, "Because you have found favor, I will explain this to you. The Day of Judgment is a day of truth. Just as no father can send his son to take his place in this world, and no son can take his father's place, so on that day, no one can plead for another. Each person will bear their own righteousness or guilt."

I asked, "But Lord, in the past, didn't Abraham pray for the people of Sodom? Didn't Moses plead for our ancestors when they sinned in the wilderness? Didn't Joshua intercede for Israel after Achan's sin? Didn't Samuel pray for the people during Saul's rule? Didn't David pray to stop the plague? Didn't Solomon pray for those worshiping in the temple? Didn't Elijah pray for rain and for the dead to rise? Didn't Hezekiah pray during Sennacherib's attack? Many righteous people have prayed for others in times of great corruption. Why won't this happen in the future?"

He replied, "This world is not the final one. That is why the strong prayed for the weak. But on the Day of Judgment, everything will change. That day will mark the end of this world and the beginning of immortality. In that world, corruption will be gone, reckless behavior will end, disbelief will disappear, righteousness will shine, and truth will stand firm. On that day, no one will be able to help those who are condemned, and no one will be able to harm those who are saved."

I said, "Then here is my final thought: It would have been better if Adam had never been created. Or if he had to exist, that he had never sinned. What is the point of life if it only leads to misery and punishment after death?

O Adam, what have you done? Your sin did not only harm you—it brought suffering to all of us who came after you. Why give us the promise of eternal life if sin leads us to death? Why speak of everlasting joy if we are trapped in failure? Why offer paradise with fruit that never spoils and endless healing if we can never reach it because of our sins?

How can those who lived righteously shine like the stars while the rest of us remain in darkness? While we lived, we never truly thought about the suffering that follows death."

He replied, "This is the struggle of humanity: Those who fail will face the torment you described, but those who overcome will receive the rewards I have promised. This is what Moses told the people: 'Choose life so that you may live.' But they did not listen to him, just as they ignored the prophets after him, and even now, they do not listen to me. That is why their destruction will not be mourned, but there will be great joy for those who are saved."

I said, "Lord, I understand now why the Most High is called merciful. He shows kindness even to those who have not yet been born. He is compassionate to those who turn to Him. He is patient with sinners. He is generous, giving freely rather than taking away. He is forgiving, offering mercy to those who repent. If He were not merciful, the world could not exist as it does now. If He were not forgiving, no one would be left alive."

Chapter 8

The Most High created this world for many people, but the next world will be for only a few. Let me explain, Esdras. If you ask the earth, it will tell you that it produces a lot of clay to make pots and jars, but only a small amount of dust from which gold is formed. The same is true for people—many are born, but only a few will be saved.

I said, "Fill my heart with understanding, and let wisdom take root in my soul. I did not choose to be born, and I will not choose when I die. My time here is short. Lord of all, let me pray before You. Give us knowledge and wisdom so our hearts can understand. Let wisdom grow in us so that even flawed people like us, who share the same human nature, may have life.

You alone exist, and we are the work of Your hands, just as You have said. You give life to our bodies, forming us in the womb. You shape our features, preserve us in fire and water, and keep us in the womb for nine months until we are born. You have provided a way for us to grow strong, giving nourishment through a mother's milk. In Your kindness, You guide us.

You teach us righteousness, lead us through Your law, and correct us when we go the wrong way. You bring us to death, but You also give us life again because we are Your creation. But if You suddenly destroy those You have shaped with such care, what was the purpose of making them in the first place?

Now I will speak. You know everything about humanity, but I mourn for Your people—Your chosen ones, Israel, the descendants of Jacob. I grieve for their suffering. I pray for myself and for them because I see the sins of the world, but I also know how quickly Your judgment will come. Please hear my prayer and listen to my words as I plead before You."

Esdras began to pray and was lifted up, saying, "O Lord, who lives forever, who rules from the highest places, whose throne is beyond all understanding, and whose glory is beyond measure. The vast armies of angels tremble before You. At Your command, they change into wind and fire. Your words never change, and Your laws stand forever. Your power dries up the ocean depths, and Your anger melts mountains. Your truth endures forever—

Listen, Lord, to the prayer of Your servant. Hear the plea of Your creation. Pay attention to my words. As long as I live, I will speak, and as long as I have breath, I will cry out to You.

Do not focus on the sins of Your people, but remember those who have faithfully served You. Do not dwell on the wrongs of those who have turned away, but think of those who have kept Your covenant despite hardships. Do not remember the wickedness of those who have lived in rebellion, but consider those who fear You and follow Your ways.

Do not let Your will be to destroy those who act like wild animals, but look instead at those who have taught Your law with wisdom. Do not be angry with those who have fallen so low, but show love to those who trust in Your greatness. We and our ancestors have lived sinful lives, but You are called merciful because You have shown mercy even to sinners. If You are willing to show kindness to those who have no good deeds to offer, then You truly are merciful.

The righteous, who have done many good works, will be rewarded for their faithfulness. But what is humanity that You should be so angry with them? What is this fragile and imperfect race that Your wrath should be upon them? The truth is, no one born into this world is without sin, and no one who has lived has been completely blameless. In this, Lord, Your justice and mercy are revealed."

"Your mercy will shine if You show compassion to those who have nothing to offer."

Then He said to me, "Some of what you have spoken is correct, and it will happen as you have said. I will not focus on those who have sinned, nor on their destruction. Instead, I take joy in the righteous—their journey, their salvation, and the rewards they will receive. So it will be, just as I have said.

Think of it this way: A farmer plants many seeds in the ground and grows many trees, but not everything he plants grows properly, and not all the trees take root. In the same way, not everyone who is born into this world will be saved."

I said, "If I have found favor with You, let me ask one more thing. If a farmer's seeds do not grow because they don't receive enough rain, or if too much rain destroys them, then they perish. In the same way, humanity, which You created with Your own hands, made in Your image, and for whom You created all things, is like that seed.

Please, do not be angry with us. Show mercy to Your people. Spare those You have chosen, for You are merciful to what You have created."

He replied, "The things of this world are for those who live now, and the things of the future are for those who will live later. You cannot love my creation more than I do, so do not compare yourself to the wicked. Still, I see that you are sincere, and I admire you for humbling yourself rather than seeking glory by comparing yourself to the righteous.

In the last days, great troubles will come upon the world because of people's pride. But listen carefully—for you and for those who seek to understand the future prepared for people like you:

Paradise is open to you. The tree of life has been planted. The future is prepared. Abundance is ready. A city has been built. Rest is available. Goodness has been perfected. Wisdom has been fully established. The source of evil has been sealed away. Weakness has been removed. Death has been hidden. Hell and corruption are forgotten. Sorrow has vanished. In the end, the treasure of immortality will be revealed.

So stop asking me about the many who will perish. When they had the chance to choose, they rejected the Most High. They refused His law and turned away from His ways. They rejected what is right and convinced themselves there was no God, even though they knew they would die.

As surely as the promises I have made to you will come true, so too will the suffering that is prepared for them.

The Most High did not create people to be destroyed. But those He made rejected Him, dishonored His name, and were ungrateful to the One who gave them life. That is why judgment is coming now.

I have not revealed these things to everyone—only to you and a few others like you."

Then I said, "Lord, You have shown me the many wonders that will come in the last days, but You have not told me when these things will happen."

Chapter 9

He said to me, "Think carefully about this. When you start seeing the signs I told you about before, you will know that the Most High is about to visit the world He created. When there are earthquakes, people in turmoil, nations falling apart, leaders unsure of what to do, and rulers confused, then you will understand that the Most High spoke of these things long ago.

Just as everything in the world has a beginning and an end, the times of the Most High follow the same pattern. The beginning is marked by wonders and great works, and the end will be revealed through signs and events. Those who are saved and escape the dangers I have described will be those who held onto their faith and lived righteously. They will see my salvation in the land I set apart long ago.

But those who ignored my ways will be filled with shock and regret. Those who rejected and mocked my commandments will suffer. Many have enjoyed blessings in life but never sought to know me. Many ignored my law, even though they had freedom and many chances to change. They refused to understand, rejecting every opportunity I gave them. But after death, when they are in torment, they will finally realize the truth.

So stop focusing on how the wicked will be punished. Instead, think about how the righteous will be saved, because this world was made for them."

I replied, "I have said this before, and I will say it again now and in the future: far more people will perish than those who will be saved, like a huge ocean wave compared to a single drop of water."

He said, "As the land is, so is the seed that is planted in it. As the flowers are, so are their colors. As the work is, so is the judgment upon it. As the farmer is, so is his harvest. Long ago, before people were even created, I prepared the world for those who would live in it. At that time, no one opposed me because no one existed yet. But now, the people I created to live in this well-prepared world—a world with everything they need and a law beyond human wisdom—have chosen corruption.

I looked at my world, and I saw that it was ruined. I looked at my earth, and I saw that wickedness had spread across it. Seeing this, I decided to spare only a few. Out of a whole vineyard, I saved just one perfect grape. Out of an entire forest, I kept only one healthy plant. Let the rest, who were born without purpose, perish. But let my chosen ones survive, for I have shaped them with great care."

"If you are willing to wait seven more days, do not fast this time. Instead, go to a field filled with flowers where no houses have been built. Eat only the plants that grow there—no meat or wine—and pray constantly to the Most High. Then I will come and speak with you again."

So I went to the field called Ardat, just as he told me. I stayed there among the flowers, eating only the plants that grew in the field. They gave me strength.

After seven days passed, I lay down on the grass. My heart became troubled again, just as it had before. My mouth opened, and I began to speak to the Most High:

"O Lord, you revealed yourself to our ancestors in the wilderness when they left Egypt and entered a land without people or food. You said, 'Listen to me, Israel. Pay attention to my words, descendants of Jacob. I have planted my law within you, and it will grow and bear fruit, bringing you glory forever.'

But our ancestors, who received your law, did not obey it. They rejected your instructions. Even so, your law has not perished, for it is yours and cannot die. But those who received it perished because they did not keep the seed of wisdom planted in them.

Think about this: when soil receives a seed, when the sea carries a ship, or when a container is filled with food or drink, if the seed, the ship, or the contents are destroyed, the container remains. But with us, it is different.

We, who have received your law, perish because of sin, along with the hearts that received it. Yet your law does not perish—it remains in its glory."

As I was thinking about these things, I looked up and saw a woman to my right. She was crying loudly, filled with deep sorrow. Her clothes were torn, and she had ashes on her head.

I put aside my own thoughts and turned to her. "Why are you crying? What is troubling you so much?"

She answered, "Leave me alone, my lord, so I may weep for myself. My sorrow is great, and my heart is broken."

I said, "Please, tell me what has happened to you."

She replied, "I was barren and had no children, even though I was married for thirty years. Every day of those thirty years, I prayed to the Most High without stopping, day and night. After thirty years, God finally heard my prayer. He saw my suffering and answered me by giving me a son. I was overjoyed, and so was my husband. Our neighbors also celebrated with us, and we all praised the Mighty One.

I raised my son with great care and love. But when he grew up, and I was preparing for his marriage, I arranged a great feast to celebrate the occasion."

Chapter 10

When my son entered the wedding chamber, he suddenly collapsed and died. We put out the lamps, and my neighbors came to comfort me. I stayed silent until the evening of the second day. When they stopped trying to console me and encouraged me to find peace, I got up in the night, left the city, and came to this field. I decided that I would stay here, refusing to eat or drink, mourning and fasting until I die.

I set aside my own thoughts and responded to her with frustration. "You are acting foolishly. Don't you see that grief surrounds all of us? Zion, the mother of us all, is mourning and has been humiliated. This is a time for sorrow, not just for you but for everyone. Yet, you cry for only one son. Look at the earth—it has lost more than you ever could. It has given life to so many people, only for most of them to be destroyed.

Who has more reason to mourn—you for your one son, or the earth, which has lost countless lives? If you say, 'But my loss is special because I carried my son with pain and suffering,' then remember, the earth also brings forth its children—humanity—and has done so since the beginning, only to give them back to the Creator.

Be strong and endure this pain. If you accept God's judgment as fair, you will see your son again in time, and you will be honored among women. Now, go back to the city and to your husband."

She replied, "I will not return to the city. I will stay here and die."

I continued speaking to her. "Don't do this. Instead, take comfort in Zion's suffering and find hope in Jerusalem's pain. Look at what has happened to us—our sacred temple has been destroyed, our altar torn down, and our place of worship left in ruins. The sound of music is gone, our songs have stopped, and the light of the lampstand has been extinguished. The ark of the covenant has been taken, our holy items have been defiled, and even the name we were known by has been disgraced.

Our free people have been taken as slaves, our priests have been burned alive, and our Levites have been captured. Our young women have been violated, our wives abused, our righteous men carried away, our children betrayed, our young men enslaved, and our strongest warriors left weak. Worst of all, the glory of Zion has been stripped away and given to those who hate us.

So let go of your grief and set aside your sorrow. If you do, the Mighty One will show you mercy again, and the Most High will bring you peace and comfort."

As I spoke to her, her face suddenly began to glow like lightning, and she became dazzling. I was filled with fear, unsure of what was happening. Then she let out a loud and terrifying cry, and the ground shook beneath her voice. When I looked again, she was gone, and in her place, I saw a city being built, with strong and magnificent foundations.

I was afraid and cried out, "Where is Uriel, the angel who spoke to me before? He led me into this great vision, but now my spirit is overwhelmed, and I feel unworthy even to pray."

As I spoke, the angel who had appeared to me earlier returned. He saw me lying on the ground like someone who was dead, completely drained of understanding. He took my hand, comforted me, and helped me to my feet. Then he said, "Why are you troubled? What is upsetting you?"

I answered, "Because you left me. I followed your instructions and went into the field, but now I have seen things I cannot comprehend."

He said, "Stand up and be strong, and I will explain what you have seen."

I replied, "Speak, my lord, but do not leave me again, or I will die before my time. I have seen things I never imagined and heard things I do not understand. Am I being deceived, or am I dreaming? Please explain this vision to me."

He answered, "Listen carefully, and I will tell you what you need to know. The Most High has revealed many hidden things to you. He has seen your deep sorrow for your people and your grief for Zion. This is what your vision means.

The woman you saw crying—the one you tried to comfort—has now disappeared. In her place, you saw a city being built. The woman represents Zion, and the city you saw is her future.

She told you about the death of her son, which symbolizes the destruction of Jerusalem. Her barrenness for thirty years represents the three thousand years before offerings were made there. After those years, Solomon built the city and began sacrifices—this is when the barren woman gave birth to her son. She raised him with great care, representing the city of Jerusalem and its people. But when her son died on his wedding day, this symbolized the fall of Jerusalem.

You saw her mourning for her son, and you tried to comfort her. These things were shown to you so you could understand them. The Most High, seeing your compassion and grief, has now revealed to you Zion's future glory and beauty.

That is why I told you to stay in the field, away from human-built houses. The Most High was going to reveal this vision to you, and no man-made structure can stand where the city of the Most High is meant to appear.

Do not be afraid, and do not let your heart be troubled. Look at the city's beauty and greatness as much as your eyes can take in. Listen to everything your ears can understand. You are blessed more than many and have been chosen by name to stand with the Most High, along with only a few others.

Tomorrow night, remain here again, and the Most High will reveal dreams to you about what will happen to the people of the earth in the last days."

So I stayed in the field and rested for another night, just as he had told me.

Chapter 11

On the second night, I had a dream. In my dream, I saw a huge eagle rise from the sea. It had twelve large wings and three heads. The eagle stretched its wings over the entire earth, and the winds of heaven blew upon it as dark clouds gathered around it. Near the larger wings, smaller wings began to grow, forming tiny ones. The three heads of the eagle remained still, but the middle head was the largest, though it did not move.

The eagle soared through the sky and ruled over the earth and everyone living under it. No creature dared to challenge its power. Then the eagle stood on its talons and spoke to its wings, saying, "Do not all watch at once. Each of you must take turns to rest and to watch in your appointed time. The heads will be saved for the final moment."

The voice did not come from the eagle's heads but from the middle of its body. I counted the small wings and saw that there were eight. Then I saw one wing on the right side rise up and begin to rule over the earth. But when its rule ended, it disappeared without a trace. Another wing rose after it and ruled for a long time, but eventually, it also came to an end and vanished, just like the first.

Then a voice spoke to this second wing, saying, "Listen, you who have ruled for so long. Before you disappear, know this: no ruler after you will reign as long as you have—not even half as long."

The third wing took power like the others before it, but it too vanished. One by one, each wing took its turn ruling, and then disappeared. The remaining wings on the right side also rose to power. Some ruled for a short time before vanishing, while others rose but never ruled at all. Eventually, all twelve wings were gone, along with two of the small wings. The only parts of the eagle left were the three resting heads and six of the smaller wings.

I then saw two of the smaller wings detach from the remaining six and move under the head on the right side, while the other four stayed in place. These two wings tried to rise and take power. One of them started to rule but soon vanished. Another rose, but disappeared even faster than the first.

The last two small wings also prepared to take power. As they hesitated, the middle head—the largest of the three—finally woke up. It joined the other two heads, and together, they acted as one. The middle head turned and devoured the two small wings that had tried to rule. Then this middle head took control of the entire earth, ruling the people with great cruelty. It was more powerful than all the wings before it. But after some time, the middle head also disappeared, just like the wings had.

The two remaining heads then ruled over the earth and its people.

Eventually, the head on the right side turned and devoured the one on the left.

Then I heard a voice say, "Look in front of you and think carefully about what you see."

I looked and saw a powerful lion come out of the forest. It roared loudly, and then it spoke to the eagle in a human voice, saying:

"Listen to the message from the Most High! Are you not the last of the four great beasts I created to rule the earth before the end of time? This fourth beast has been the most terrifying of them all. You have ruled the world through fear and cruelty, and your time has lasted longer than the others. But you have ruled with lies and deception.

You have judged the world unfairly. You have hurt the weak, mistreated those who live in peace, hated those who speak the truth, and loved liars. You destroyed the homes of good people and tore down the walls of those who never wronged you. Your arrogance has reached the Most High, and your pride has risen before the Mighty One.

The Most High has examined the times, and they are now complete. The days of your rule are over.

So, eagle, you will be seen no more—your cruel wings, your wicked smaller wings, your ruthless heads, your sharp talons, and your corrupt body will all be gone. The earth will finally have peace and hope. It will be freed from your violence and will look forward to the judgment and mercy of its Creator."

Chapter 12

As the lion spoke to the eagle, I watched as the last head of the eagle disappeared. The two wings that had moved to support it took power, but their rule was short and filled with chaos. Soon, they also vanished, and the entire body of the eagle was consumed by fire. The whole earth trembled in fear.

I woke up deeply shaken, my mind troubled and my heart heavy with dread. I said to myself, "Seeking to understand the ways of the Most High has overwhelmed me. My mind is exhausted, and my spirit feels weak. The fear from this vision has drained my strength. I must ask the Most High to help me endure." Then I prayed, "O Lord, if I have found favor in Your sight, if I stand before You as righteous, and if my prayer has reached You, please strengthen me. Reveal to me the meaning of this terrifying vision so that my soul may find peace. You have judged me worthy to see the end times and the events that will unfold in the final days."

Then the Most High spoke to me, saying, "This is what your vision means. The eagle that rose from the sea represents the fourth kingdom, the same one I showed to your brother Daniel in his vision. But now, I will give you an explanation that was not fully revealed to him.

A time will come when a powerful kingdom will rise on the earth, greater and more feared than any that came before it. It will be ruled by twelve kings, one after another. Of these, the second king will rule the longest.

This is what the twelve wings you saw represent. As for the voice that came from the middle of the eagle's body, not from its heads, it means that after this kingdom has ruled for a time, internal conflicts will threaten to destroy it. However, it will not completely fall but will regain its strength for a little while.

The eight smaller wings attached to the larger ones represent eight kings who will rise from within this kingdom. Their rule will be brief, and they will pass quickly. Two of them will die as the kingdom reaches its middle years. Four will survive until the final days, and two will remain until the very end.

The three resting heads symbolize the last three rulers of this kingdom. The Most High will allow them to rise, and they will bring great oppression to the world, worse than anything seen before. This is why they were shown as the eagle's heads. These three rulers will complete the wickedness of the eagle and carry out its final deeds.

The disappearance of the largest head means that one of these rulers will die in his bed, but he will suffer greatly before his death. As for the two remaining heads, they will be killed by the sword. One will turn against the other and destroy it, but the one that remains will also be struck down in the final days."

"The two smaller wings that moved over to the head on the right side represent those whom the Most High has set apart for the last days. Their rule will be short and filled with unrest. The lion that came from the forest, roaring and speaking to the eagle, is the chosen one of the Most High. This anointed ruler, a descendant of David, will confront the eagle for its evil, corruption, and arrogance. He will judge these rulers while they are still alive, and after delivering his judgment, he will destroy them.

Afterward, he will show mercy to the people who remain—those who were preserved within My borders. He will bring them joy until the time of the final judgment, the very judgment I spoke to you about from the beginning."

"This is the meaning of the vision you saw. Only you have been chosen to understand these mysteries. Write down everything you have seen in a book and store it safely. Share it only with the wise among your people—those who have the heart to understand and protect these secrets. For now, remain here for seven more days. The Most High will reveal even more to you."

With that, He left me.

After seven days had passed, the people—both the lowly and the powerful—realized that I had not returned to the city. They gathered together and came looking for me.

They said, "What have we done to offend you? Why have you abandoned us and stayed here alone? You are the only prophet left to guide us. You are like the last cluster of grapes after the harvest, a lamp shining in the darkness, or a safe harbor for a ship that has survived a storm. Haven't we suffered enough? If you leave us now, it would have been better if we had died in the flames of Zion. We are no better than those who perished there."

Then they began to weep loudly.

I replied, "Take heart, Israel! Do not lose hope, descendants of Jacob. The Most High has not forgotten you. The Mighty One always remembers His people. As for me, I have not abandoned you. I came here to pray for Zion, to plead for mercy over the destruction of the holy place. Now, return to your homes. When my time here is finished, I will come back to you."

After hearing my words, the people returned to the city as I instructed.

I remained in the field for seven days, just as the angel had commanded me. During this time, I ate only the flowers of the field, and my nourishment came from the plants.

Chapter 13

After seven days, I had a dream at night. In my vision, I saw a powerful wind rise from the sea, stirring its waves violently. From the middle of the sea, something appeared that looked like a man. He flew on the clouds of heaven, and whenever he turned to look, everything before him trembled with fear. When he spoke, his voice was so powerful that people who heard it melted away like wax in a fire.

Then I saw a huge crowd of people gathering from all directions under the sky. They came together to fight against the man who had risen from the sea. The man carved out a massive mountain for himself and flew to stand on it. I tried to see where the mountain had come from, but I couldn't tell. Even though the great army prepared to attack him, they were filled with fear. Still, they dared to fight him.

As they moved against him, the man didn't raise a hand or use any weapon. Instead, a fiery stream came from his mouth, a strong wind from his lips, and a storm of sparks from his tongue. These forces combined into one powerful blast, striking the attacking army and destroying them in an instant. Suddenly, the countless soldiers turned into nothing but dust and ashes, and the air was filled with smoke. Seeing this, I was completely in awe.

After this, the man came down from the mountain and called another group of people to himself, but this group was peaceful. Many approached him—some were joyful, others were sorrowful. Some were in chains, while others brought gifts. Overwhelmed with fear, I woke up and prayed to the Most High, saying, "Lord, You have shown me incredible wonders and have chosen me to receive this vision. Now, I beg You, please explain its meaning to me.

I know that those who survive in those days will go through great suffering. But those who do not survive will also suffer, tormented by their understanding of what is to come. Woe to those who remain, for they will face danger and hardship just as I saw in my dream. Still, it is better to go through those trials and see the end than to vanish like mist, never witnessing the events of the last days."

Then the Most High answered me, "I will explain the meaning of the vision and answer your questions.

You asked about those who will survive. Here is the answer: Those who make it through the dangers of those days will help others who are also struggling. These people have shown faith and lived righteously in My sight. Know this—those who remain are more blessed than those who perish.

Now, about your vision: The man rising from the sea is the one the Most High has been keeping for many ages. He will deliver My creation by his own power and guide those who survive.

The wind, fire, and storm coming from his mouth represent how he will destroy the armies that come against him, without using physical weapons. The time is coming when the Most High will begin to save those who live on the earth. When that happens, fear and confusion will spread among the people. Cities will turn against each other, nations will be in chaos, and kingdoms will collapse. When these events begin, and the signs I have shown you take place, then My Son—the one you saw as a man rising—will be revealed.

When the nations hear his voice, they will stop fighting each other and instead unite to fight against him. A massive army will form, just as you saw in your vision. My Son will stand on Mount Zion, which will be revealed to all people, fully prepared and established, just as you saw the mountain being carved out without human hands. He will confront these nations because of their wickedness and bring plagues upon them like a violent storm, exposing their evil actions and thoughts. Without any effort, he will destroy them with the law, which is as powerful as fire.

The peaceful crowd that gathered to him represents the ten tribes of Israel. Long ago, during the reign of King Hosea, these tribes were taken from their homeland and exiled by King Shalmaneser of Assyria. They were led to a distant land across the river. Among themselves, they decided to leave the company of other nations and journey to an uninhabited region where they could follow My laws, which they had failed to obey in their own land.

They crossed the narrow passages of the Euphrates River, and the Most High performed miracles for them, stopping the river's flow until they had safely crossed. Their journey through that region lasted a year and a half, and the land they reached is called Arzareth. They have lived there until the end times. But when it is time for them to return, the Most High will once again stop the river so they can cross. This is the peaceful group you saw gathered together.

The survivors from among your people are those within My holy borders. When My Son destroys the gathered nations, he will protect the remaining people and reveal many great wonders to them."

Then I asked, "Lord, why did the man appear from the sea?"

He answered, "Just as no one can fully explore or understand the depths of the sea, no one on earth can see My Son or those with him until the time of his appearance.

This is the meaning of your vision, and it has been revealed only to you because you have set aside your own ways to follow Mine. You have committed yourself to My law and shaped your life according to wisdom, making understanding your guide. That is why I have shown you these things. A reward is prepared for you with the Most High. After three more days, I will reveal even greater and more amazing things to you."

I left and went into the field, praising and giving thanks to the Most High for His wonders. He works everything at the right time, and He rules over all things according to their proper seasons. I stayed there for three days.

Chapter 14

On the third day, as I sat under an oak tree, a voice called to me from a nearby bush, saying, "Esdras, Esdras!" I answered, "Here I am, Lord," and stood up. The voice continued, "I appeared in a burning bush and spoke to Moses when my people were slaves in Egypt. I sent him to free them and led him to Mount Sinai, where he stayed with me for many days. There, I showed him amazing things, revealed secrets about time and the end of the world, and gave him this command: 'Some things you will share, but others you must keep hidden.'

Now, I am telling you the same. Keep in your heart the signs you have seen, the dreams you have had, and the meanings you have learned. Soon, you will leave this world and join my Son and others like you until the end of time. The world is growing weaker, and time is running out. The age has been divided into twelve parts—ten and a half have already passed, and only two and a half remain.

So now, set your affairs in order. Warn your people, comfort those who are humble, and guide the wise. Let go of this temporary life and free yourself from the concerns of the flesh. Put aside your human worries and do not cling to weakness. Let go of your greatest fears and focus on preparing for what is coming. For the troubles you have already seen are nothing compared to what is yet to come. As the world ages, its evils will increase. Truth will become rare, and lies will grow stronger. The eagle you saw in your vision is rushing toward its destined time."

I answered, "Lord, please allow me to speak. I will warn the people alive today, as you have commanded. But who will warn the future generations? The world is covered in darkness, and the people in it have no light. Your law has been destroyed, and no one remembers your works or the things you have planned. If I have found favor with you, send your Holy Spirit to me so I can write everything down—from the beginning of time until now, including your law—so that future generations may have guidance and hope in the last days."

He answered, "Go and gather the people. Tell them not to look for you for forty days. Prepare many writing tablets and bring five men skilled in writing: Sarea, Dabria, Selemia, Ethanus, and Asiel. Come to this place, and I will fill your heart with understanding that will not fail until you have written everything I tell you. When you are finished, some books will be made public so that everyone can read them, but others you will keep secret and share only with the wise among your people. Tomorrow at this time, you will begin writing."

I went and gathered the people as I was told. I said, "Listen to me, O Israel! Our ancestors were strangers in Egypt, but they were rescued and given the law of life. Yet, they did not obey it, and neither have you. The land of Zion was given to you as an inheritance, but you and your ancestors acted wickedly and disobeyed the Most High. Because He is a just judge, He has taken away what He once gave you.

Now, you are gathered here with your families. If you take control of your thoughts and guide your hearts wisely, you will preserve your lives. And after death, you will receive mercy. For after death comes judgment, and then we will live again. At that time, the names of the righteous will be revealed, and the sins of the wicked will be exposed. But for now, do not seek me or come to me for forty days."

I took the five men as instructed and went into the field, where we remained. The next day, a voice called to me, saying, "Esdras, open your mouth and drink what I give you." I opened my mouth, and a cup was handed to me. It looked like it was filled with water, but it shone like fire. I drank from it, and my heart was filled with understanding. Wisdom grew within me, and my spirit remembered everything. My mouth was opened, and I began to speak without stopping.

The Most High gave understanding to the five men, and they wrote down everything I spoke in a language they did not know. They wrote all day and ate bread only at night. I continued speaking, even through the night, without resting. By the end of the forty days, ninety-four books had been written.

When the forty days were completed, the Most High said to me, "Make the first books public so that both the worthy and the unworthy may read them. But keep the last seventy books and share them only with the wise among your people. These books contain the fountain of understanding, the source of wisdom, and the stream of knowledge." I did as He commanded.

Chapter 15

"Give this message to my people—the words I will tell you," says the Lord. "Write them down because they are true and reliable. Do not be afraid of those who plot against you, and do not let the disbelief of others disturb you. Those who refuse to believe will be destroyed because of their unbelief.

Look," says the Lord, "I am bringing disasters across the entire earth—war, famine, death, and destruction. Wickedness has spread through every nation, and their evil has reached its limit. Because of this," declares the Lord, "I will no longer remain silent about their sins, nor will I ignore their wrongdoing any longer. The cries of the innocent and righteous, whose blood has been spilled unfairly, reach me constantly. I will surely take revenge on their behalf," says the Lord, "and I will demand justice for all the innocent blood that has been shed.

Look, my people are being led like sheep to the slaughter. I will no longer allow them to remain in Egypt. I will bring them out with a mighty hand and a powerful arm. I will strike Egypt with plagues, just as I did before, and I will make the land desolate. Let Egypt and its people mourn, for I will bring disaster upon them. Let the farmers grieve, for their crops will fail, and their trees will be ruined by disease, hail, and powerful storms.

Trouble is coming for the whole world and everyone who lives in it! Destruction and war are near. Nations will rise against each other, and people will take up weapons against their own neighbors. There will be no respect for leaders or authorities. People will try to enter cities but will not be able to. Pride and arrogance will tear cities apart, homes will be destroyed, and fear will grip the people. No one will care for their neighbor. Instead, they will invade each other's homes with weapons, stealing out of desperation due to hunger and hardship.

Watch," says God, "as I summon all the kings of the earth—from the east, the south, the north, and from Lebanon. They will turn against each other, and I will repay them for what they have done to my people. Just as they have mistreated my chosen ones, I will repay them in full," says the Lord God. "I will not spare the wicked, and my sword will not rest until justice is served for the innocent. My wrath burns like a fire, consuming sinners like dry straw.

Woe to those who live in sin and refuse to obey my commands!" declares the Lord. "I will not have mercy on them. Depart from me, you rebels! Do not defile my sanctuary!" The Lord knows all who have sinned against Him, and He will hand them over to destruction.

Disaster has already begun to spread across the earth, and you will not escape it because you have rejected God.

Look, a terrifying army is coming from the east! The warriors of Arabia, like fierce dragons, will march with many chariots. Their battle cries will echo across the land, striking fear in all who hear them. The warriors of Carmonia, raging like wild boars, will charge with great power, crushing everything in their path as they attack the land of Assyria. The dragons will overpower their enemies with great strength, but their foes will regroup and strike back, pushing them into retreat.

From Assyria, an army will rise and ambush them, crushing their forces. The soldiers will be gripped with fear, and their leaders will be thrown into confusion. Then, thick clouds will come from the east, north, and south, filled with terrifying storms. These clouds will crash into one another, creating a great storm over the earth. Blood will flood the land, rising as high as a horse's belly, a man's thigh, and a camel's knee. Fear and panic will spread across the world, and those who witness this destruction will tremble in terror.

After this, violent storms will come from every direction—north, south, and west. But a powerful wind from the east will drive the storm forward with unstoppable force. Huge clouds filled with fury will rise, bringing disaster to the land and its people. These storms will rain down fire, hail, flying swords, and massive floods. Rivers and valleys will overflow, and every low-lying area will be covered in water. The storms will destroy cities, tear down walls, flatten mountains, and sweep away forests, fields, and crops.

The storm will press on toward Babylon, surrounding the city and unleashing destruction. Thick smoke and dust will rise into the sky, and those who see Babylon's downfall will cry in sorrow. The few survivors will become slaves to their conquerors.

Woe to you, Asia! You took pride in Babylon's riches and followed in her ways. You dressed your daughters like prostitutes to attract lovers, just as Babylon did. You copied her wickedness and her deceitful schemes. Because of this," says God, "I will bring disaster upon you: widowhood, poverty, famine, war, and disease. These will devastate your homes and bring destruction upon you.

Your strength and beauty will fade like a flower under scorching heat. You will be weak, like a woman beaten down, unable to defend yourself or those you depended on. Do you think I would be so angry with you," says the Lord, "if you had not continually attacked my chosen people? You celebrated their deaths, mocked their suffering, and took joy in their destruction.

Go ahead, put on your finest clothes! But you will receive the wages of a prostitute—your payment is coming. Just as you have treated my people," says the Lord, "so will I treat you. You will be handed over to your enemies.

Your children will die from hunger. You will be struck down by the sword. Your cities will be ruined, and those who flee to the countryside will not survive. Those who hide in the mountains will starve, resorting to eating their own flesh and drinking their own blood in desperation. You, the most wretched of all, will face even greater horrors. Armies will march against your land, bringing destruction and wiping away your pride. Then, they will return to the ruins of Babylon.

They will burn you like straw in a raging fire. Your cities, your land, your mountains, your forests, and your fruit trees will all be turned to ash. They will take your children as slaves, steal your treasures, and destroy your beauty."

Chapter 16

Woe to you, Babylon and Asia! Woe to you, Egypt and Syria! Wear sackcloth and mourn, for your time of destruction is near. Cry for your children and grieve, for disaster is coming. A sword has been sent against you—who can turn it back? Fire has been unleashed—who can put it out? Trouble is on the way—who can stop it? Can a starving lion be forced back into the forest? Can a wildfire in dry grass be put out once it has started? Can an arrow shot by a skilled archer be brought back?

The Lord God has sent these disasters, and no one can stop them. His fire burns fiercely, and no one can put it out. His lightning will strike, and everyone will tremble in fear. His voice will thunder, and no one will escape His judgment.

The earth will shake, and its very foundations will tremble. The sea will roar and crash with towering waves, terrifying even the creatures in the water. This will happen when the Lord reveals His power. His mighty hand will pull back the bow, and His arrows will hit their mark without fail. His disasters will not be stopped or turned away. The fire He has started will burn until the foundations of the earth are consumed. Just as an arrow does not return to its bow once it is shot, these disasters will not stop until their purpose is fulfilled.

Woe to me! Woe to me! Who will save me in those days? Terrifying times are coming. There will be sorrow everywhere. Famine will spread, and countless will die. Wars will erupt, and even the strongest will be filled with fear. Disaster after disaster will come, and every heart will tremble. What will people do when these troubles strike? Hunger, disease, and suffering will come to correct them, yet they will refuse to change. They will not turn away from their sins or recognize why they are being punished.

Food will become so cheap that people will think everything is fine.

But then, destruction will strike even harder—wars will spread, famine will worsen, and chaos will cover the land. Many will die from starvation, and those who survive will be killed in battle. The dead will be left unburied, and there will be no one to comfort the living. The land will be empty, and cities will be ruined. Farmers will no longer work their fields, and no one will plant crops. Trees will still bear fruit, but there will be no one left to harvest them. Grapes will grow, but no one will press them into wine. The land will be silent and deserted.

People will search for others, but only a few will be left. In a city, maybe ten people will remain. In the countryside, only two will survive, hiding in forests or caves. It will be like an olive tree that has only a few olives left after harvest or a vineyard with just a handful of grapes after being picked. Only a small number of people will be left, while soldiers move from house to house with their swords.

The land will become wild and overgrown. Weeds will take over fields, and roads will be covered with thorns because no one will travel them anymore. Young women will mourn because there will be no bridegrooms. Wives will grieve for their husbands. Daughters will cry because they have no one to protect them. Their husbands and bridegrooms will die in wars, and others will starve to death.

Listen carefully, all who serve the Lord, and understand these words. Pay attention to what God is saying, and do not doubt His warnings. These disasters are coming soon, and they will not be delayed.

Just as a woman feels labor pains before giving birth, and the child arrives quickly, these disasters will come suddenly upon the earth. The world will cry out in pain, and suffering will spread everywhere.

"My people, hear my words: prepare for battle and live as if you are only passing through this troubled world. If you sell something, act as if you must flee. If you buy something, do not expect to keep it. If you do business, do not assume you will profit. If you build a home, do not expect to live in it. If you plant crops, do not expect to eat them. If you tend vineyards, do not assume you will gather the grapes. If you marry, live as if you will not have children. If you are unmarried, live as though you are already widowed.

All your efforts will be in vain. Foreigners will take your crops, steal your possessions, destroy your homes, and take your children. Those who survive will suffer from famine and be taken as prisoners. Even if they have children, they will only experience more sorrow.

Merchants will trade, only to be robbed. The more people decorate their cities, homes, and belongings, the more I will despise them for their wickedness," says the Lord. "Just as a faithful woman hates the ways of a prostitute, so will righteousness despise evil.

When sin tries to make itself look good, righteousness will expose it. When the Defender of the faithful arrives, He will reveal every hidden sin. So do not follow the ways of wickedness or imitate those who do evil.

Soon, sin will be wiped from the earth, and righteousness will rule over all. No sinner can claim they have done no wrong, for God will bring judgment on those who say, 'I have never sinned before God.' The Lord knows every human thought, plan, and intention. He spoke, and the earth was created. He commanded, and the sky was formed. By His word, the stars were placed in the heavens, and He knows each one by name.

He searches the deepest parts of the sea and knows all its treasures. He has measured the oceans and set their limits. By His command, He holds the earth in place above the waters. He stretched out the heavens like a canopy and secured them over the waters. He made springs in the deserts and pools on the mountaintops, sending rivers to water the land.

He created people and gave them hearts to understand. He gave them life, breath, and the spirit of wisdom. He made everything and sees all that is hidden, even the deepest secrets. Surely, He knows your thoughts and the plans in your heart.

Woe to those who sin and try to hide it! The Lord will carefully examine every action and expose all wrongdoing. When your sins are revealed for all to see, you will be filled with shame, and your own deeds will testify against you. What will you do then? How will you hide from God and His angels?

Look, God is the judge. Fear Him! Turn away from sin, leave behind your wickedness, and never return to it. Then God will guide you and rescue you from suffering.

Look, a furious army is coming, filled with anger. They will capture some of you and force you to eat food offered to idols. Those who refuse will be mocked, hated, and trampled. In the cities, there will be uprisings against those who fear the Lord. These attackers will act with reckless violence, sparing no one. They will kill and destroy those who worship God, stealing everything they own and forcing them out of their homes.

At that time, My chosen people will be tested, just as gold is purified in fire.

Listen, My chosen ones," says the Lord. "The days of trouble are near, but I will save you from them. Do not be afraid, and do not doubt, for I am your guide.

You who obey My commandments and keep My teachings," says the Lord God, "do not let your sins weigh you down. Do not let your wrongdoing pull you away from Me.

Woe to those who are weighed down by sin, trapped in wickedness like a field overrun with thorns! No one can walk through it. That field will be abandoned and burned in the fire."

Second Baruch

Yerusalem will be destroyed.

Chapter 1~5

In the twenty-fifth year of King Jeconiah's reign over Judah, the Lord spoke to Baruch, the son of Neriah, and said:

"Do you see what these people are doing to me? The sins of the remaining tribes are even worse than those committed by the ten tribes who were taken into exile. The ten tribes were led astray by their kings, but these two tribes push their kings into sin.

Because of this, I will bring disaster upon this city and its people. They will be taken away from my presence for a time, and I will scatter them among the nations. Yet through them, other nations may be blessed. My people will go through discipline, but the time will come when they will seek the things that bring them peace and prosperity.

I am telling you this so you can warn Jeremiah and others like you to leave this city. Your actions are like a pillar that holds it up, and your prayers are like a strong wall protecting it."

And I said, "Lord, my God, did I come into this world only to witness the suffering of my people? No, my Lord. If I have found favor in your eyes, take my life first, so I may join my ancestors instead of seeing the destruction of my people.

I feel trapped, caught between two forces. I cannot go against You, yet my soul cannot bear to see my people suffer.

But I have one question, Lord. What will happen after all this? If

You destroy this city and give our land to those who hate us, how will Israel ever be remembered?

Who will declare Your greatness? Who will teach others Your laws? Will the world return to chaos and silence? Will people's souls be taken away, and will the memory of humanity disappear forever?

And what about the promises You made to Moses about us?"

The Lord answered me, "This city will be taken for a time, and the people will face discipline, but the world will not be forgotten.

Do you think this city is the one I spoke about when I said, 'I have engraved you on the palms of my hands'? The city I was speaking of is not the one you see now. It is the one that will be revealed with me. It was prepared from the moment I created the Garden of Eden, and I showed it to Adam before he sinned.

When Adam disobeyed, this city and the Garden were taken from him. Later, I revealed it to Abraham during the night of the covenant, between the divided sacrifices.

I also showed it to Moses on Mount Sinai when I gave him the design for the tabernacle and all its furnishings.

Now, this city is kept safe with me, along with the Garden. Go now and do what I have commanded you."

I replied, "Then I will be guilty in Zion, for Your enemies will invade this land and desecrate Your holy place. They will take Your people as captives and rule over those You love.

They will return to their land filled with idols and brag about their victory. Lord, what will You do for Your great name?"

The Lord answered, "My name and my glory will last forever. My judgment will come at the right time.

You will see with your own eyes that the enemy will not destroy Zion or burn Jerusalem. Instead, they will serve the true Judge for a time.

But as for you, go and do all that I have commanded you."

So I gathered Jeremiah, Iddo, Seraiah, Jabesh, Gedaliah, and all the leaders of the people. I led them to the Kidron Valley and told them everything the Lord had spoken to me.

They lifted their voices and wept loudly. We sat together and fasted until evening.

Chapter 6~9

The next day, a Chaldean army surrounded Jerusalem. As evening fell, I, Baruch, stepped away from the people and left the city walls. I stood near an old oak tree, heartbroken over Zion and grieving for the people of Israel who were about to be taken into exile.

As I mourned, a powerful wind suddenly swept me off the ground. It carried me high above the walls of Jerusalem, giving me a full view of the entire city. As I looked down, I saw four angels standing at the four corners of the city, each holding a burning torch. Their flames glowed brightly, casting a strange and ominous light over Jerusalem.

Then, another angel descended from the sky, radiant and full of authority. He turned to the four angels holding the torches and said, "Hold your lamps steady and do not light them until I give the command. First, I must complete my task—delivering a message to the earth and carrying out the Lord's command."

I watched as the angel entered the Holy of Holies inside the temple. There, he gathered the most sacred objects: the veil, the holy ephod, the mercy seat, the two stone tablets of the law, the priests' holy garments, the altar of incense, the forty-eight precious stones from the priests' robes, and all the sacred vessels used in worship.

After collecting these items, the angel raised his voice and spoke to the earth:

"Earth, earth, earth, listen to the word of the Mighty God. I am entrusting you with these holy objects. Keep them safe until the appointed time. When the moment comes, you will return them so they will not fall into the hands of those who do not honor them.

For now, Jerusalem will be given over to destruction, but one day, it will be restored forever."

As soon as he finished speaking, the ground opened up and swallowed the sacred objects.

Then, the angel turned to the four holding the torches and commanded, "Now, bring down the walls of this city. Tear them to their very foundations so that our enemies cannot boast and say, 'We destroyed Zion and burned the house of the Mighty God.'"

After speaking, the angel returned me to my place near the oak tree.

Then, the angels carrying the torches began their work. They struck the four corners of the walls, and the entire structure collapsed. As the city crumbled, a voice echoed from within the temple, crying out:

"Enter now, enemies! Come forward, adversaries! The One who protected this house has now abandoned it."

With a heavy heart, I, Baruch, turned away and left.

Soon after, the Chaldean army stormed the city. They rushed into the temple, seizing everything inside and around it. They captured the people, taking many into exile, killing others, and leaving only a few survivors.

King Zedekiah was caught, bound, and taken as a prisoner to the king of Babylon.

I found Jeremiah, who was spared because of his pure heart and was not taken during the invasion. Together, we tore our clothes in sorrow and wept bitterly for Jerusalem. We mourned and fasted for seven days, grieving deeply for our people and our land.

Chapter 10~12

After seven days had passed, the word of God came to me, saying, "Tell Jeremiah to leave this place and go with the exiles to Babylon. He will encourage and support them there. But you, Baruch, must stay here in the ruins of Zion. In the coming days, I will show you what will happen at the end of time."

I did as the Lord commanded and told Jeremiah His words. Without hesitation, Jeremiah left with the people who were being taken into exile. Meanwhile, I, Baruch, returned to Jerusalem and sat in front of the gates of the Temple. Surrounded by the ruins of the holy city, I raised my voice and mourned for Zion:

"Blessed are those who were never born, or those who lived and have already passed away.

But as for us who remain, woe to us! We must witness Zion's suffering and the destruction of Jerusalem.

I call upon the sirens of the sea. Creatures of the wilderness, spirits of the desert, and monsters of the forests—awake! Come and mourn with me for the fall of Zion. Let us sing songs of sorrow together.

You, farmers, stop planting your fields. And you, earth, why do you still bring forth fruit? Hold back your harvest and keep it hidden.

You, vineyards, why do you still produce wine? No offerings will ever again be made from you in Zion, and the first fruits will never again be placed on the altar.

You, heavens, stop sending rain. Lock up the dew in your storehouses.

You, sun, stop shining. And you, moon, let your glow fade away. Why should light continue when Zion's radiance has been extinguished?

Bridegrooms, do not enter your wedding chambers. Brides, throw away your garlands.

Women, stop praying for children, for the barren will rejoice more than mothers. Those without children will find joy, but those with sons will only grieve.

Why should mothers endure the pain of childbirth, only to bury their children in sorrow? Why should fathers bring sons into a world where they will be taken into exile or killed?

From now on, do not speak of beauty or gracefulness, for they are no longer found in this land of ruin.

Priests, take the keys to the Temple and throw them toward the heavens. Say to the Lord, 'Guard Your house Yourself, for we were unworthy caretakers and have failed You.'

And you, young women who weave fine garments with gold from Ophir, gather your treasures and throw them into the fire. Let the flames carry them back to the One who created them, so our enemies do not take them for themselves."

Then I turned my sorrow toward Babylon and said,

"Babylon, if Zion had remained glorious while you prospered, it would have been unbearable to see you as her equal. But now, our grief is endless, and our sorrow has no measure. Zion is in ruins, while you thrive.

Who will judge these injustices? To whom can we cry out for fairness?

O Lord, how can You bear this? Our ancestors died without witnessing this suffering. The righteous rest in peace underground, unaware of the pain we now endure.

If only the earth had ears to hear and a heart to understand! If only the dust could send a message to the dead, telling them, 'You are more fortunate than those who are still alive.'

But I will speak the thoughts of my heart. I will cry out against you, O land of prosperity:

The sun does not shine endlessly at noon, and its light does not last forever. Do not think your wealth and power will never end. Do not be arrogant or oppress others, thinking you will always remain above them.

For judgment will rise against you at the appointed time. Though the Lord's patience holds it back for now, it will surely come."

After speaking these words, I fasted for seven days, filled with sorrow and prayer.

Chapter 13~15

After all these things happened, I, Baruch, stood on Mount Zion, looking at the ruins of the city, lost in thought. Suddenly, a voice from above called out to me:

"Stand up, Baruch, and listen to the words of the Mighty God. You are shocked by what has happened to Zion, but know this: you will be kept safe until the end of time so that you can be a witness.

When the great cities of the world ask, 'Why has God allowed such destruction?' you and others who have seen this disaster will be able to answer them, saying, 'This happened so that all nations could be judged for their actions.'

And if they ask, 'When will this judgment come?' you must say:

'You who enjoyed the sweet wine of success, now drink the bitter cup of suffering, for God's judgment is fair.

He did not even spare His own people when they sinned but punished them like enemies so they might be forgiven. But you, the nations of the world, have also done wrong. You have mistreated the earth and used its gifts selfishly.

You have taken all the good things God created, yet you have shown no gratitude for His blessings.'"

Hearing this, I said, "You have shown me the future and how judgment will come upon the nations.

But I do not understand. Many who have done evil have lived comfortable lives and died peacefully. When the time of judgment comes, few will be left to hear Your words.

What is the point of knowing this, Lord? What punishment could be worse than what we have already suffered?

Yet I will continue to ask You: What good is it for those who have lived righteously, those who have stayed away from the ways of the world and never sought answers from the dead?

These faithful ones have feared You and followed Your ways. And yet, You allowed Zion to be destroyed without mercy, even though these good people lived there.

If others have sinned, should Zion not have been spared for the sake of the righteous? Why was she destroyed because of the wicked?"

I paused, then asked, "Who, O Lord, can understand Your judgment? Who can comprehend Your ways or grasp the greatness of Your plan?

No one born on earth can fully understand the depth of Your wisdom.

We are as brief as a passing breath. Just as a breath disappears, so does human life. We do not leave this world by our own choice, and we do not know what will happen to us after we are gone.

But the righteous have hope in the end. They leave this life without fear, trusting in the good deeds they have stored with You. These are treasures kept safe in Your heavenly storehouses.

So they depart with peace and confidence, knowing that a better world awaits them.

But woe to us who remain, suffering and waiting for even more hardship. You, Lord, know everything about Your creation and why You made us. We cannot understand goodness as You, our Creator, do.

Still, I will speak. In the beginning, before the world existed, You spoke a single word, and creation came to life.

You made humanity to rule over the earth so that it would be clear that the world was created for them, not the other way around.

Yet now, I see that the world continues, but we, the ones it was made for, pass away."

Then the Lord answered me, "You are right to be amazed at how short man's life is. But your understanding of why the wicked prosper and the righteous suffer is incomplete.

You have said that people cannot understand My judgment, and that is true. But listen carefully, and I will explain it to you.

If I had not given humanity My law and taught them wisdom, they would not be able to understand My justice. But now that they have received knowledge and still choose to do wrong, they will be judged according to what they know.

As for the righteous, for whom you say the world was created: their true reward is not in this world, but in the world to come.

This life is full of hardship and struggle for them, but the next life will be their crown of glory—a reward far greater than anything this world could offer."

Chapter 16~20

I replied, "O Lord, my God, our lives are short and full of struggles. How can anyone, in such a brief and difficult life, inherit something infinite and beyond measure?"

The Lord answered me, "With the Most High, time is not measured by how long or short it is. Think about Adam—he lived for 930 years, but what good did all those years do him? His disobedience brought death upon himself and shortened the lives of all his descendants.

But look at Moses—he lived only 120 years. Yet, because he obeyed the One who created him, he was able to bring the law to the descendants of Jacob. Through his faithfulness, he became a guiding light for the people of Israel."

I then said, "The one who lit the lamp shared its light, but so few have followed it. Instead of rejoicing in the light he gave, many have chosen the darkness of Adam and refused the guidance of the lamp."

The Lord replied, "That is why I made a covenant with them long ago and declared, 'I have set before you life and death.' I called the heavens and the earth as witnesses against them. I knew their time would be short, but the heavens and the earth will endure forever.

Even after death entered the world, they continued to sin, despite having the law to guide them. They were given the light of truth without fault, yet they chose rebellion. They had witnesses—the heavens, the earth, and even Me—yet they still turned away.

Now, everything that exists is under My judgment.

But you, Baruch, do not dwell on the past or let it trouble your heart. Instead, focus on the end of time, for the conclusion is what truly matters.

Think about this: If a man enjoys prosperity in his youth but suffers disgrace in his old age, he forgets the good times. But if a man struggles when he is young but finds success in his later years, he no longer remembers his past hardships.

In the same way, even if every person had lived in endless prosperity since the day death entered the world, it would all be meaningless if destruction was waiting for them in the end."

Then the Lord continued, "Look, the days are coming when time will move faster than ever before. The seasons will pass more quickly than those that have come before, and the years will fly by even faster than they do now.

That is why I have taken Zion away—so that I may soon bring about the appointed time when I will visit the world in judgment.

Now, Baruch, hold tightly to everything I have commanded you. Keep My words deep in your heart and never forget them.

Prepare yourself, for soon I will reveal My great judgment and My mysterious ways to you.

So, set yourself apart for seven days. During this time, do not eat bread, do not drink water, and do not speak to anyone.

After these seven days, return to this place, and I will appear to you. I will reveal truths to you and show you what is to come. These things will not be delayed—they will happen just as I have planned."

Chapter 21

Then I left and went to the Kidron Valley, where I found a cave and sat inside. There, I dedicated myself to the Lord, setting my heart apart for Him. During those days, I ate no bread, yet I did not feel hungry. I drank no water, yet I did not feel thirsty. I remained in the cave for the full seven days, just as He had commanded me.

After the seven days, I got up and returned to the place where He had spoken to me before. It was sunset, and as the sky darkened, my mind became filled with deep thoughts. In that quiet moment, I began to speak before the Mighty One:

"O Lord, Creator of the earth, hear me! You, who spoke the heavens into place and set them high above by Your power. At the beginning of time, You called into existence what did not yet exist, and it obeyed Your command.

You are the one who controls the movement of the air and the forces of nature with a simple sign. You see everything—what has already happened and what is still to come. With Your great wisdom, You rule over the countless holy ones who stand before You, glowing like flames of fire, whom You created to surround Your throne.

Everything that exists belongs to You alone, and at any moment, You have the power to bring forth whatever You desire.

You count every drop of rain before it falls to the earth. You know the end of time before it even begins. Please hear my prayer, for You alone sustain all things—those who live now, those who have passed, and those yet to come.

You, O Lord, are the Living One, the Unchanging One, the Eternal One. You alone know the number of mankind. Though many throughout history have sinned, many others have walked in righteousness.

You understand the destiny of both the wicked and the righteous. You have set the final outcome of each one.

If this life is the only one we have, then nothing could be more painful than knowing how temporary it is. What good is strength if it fades into weakness? What is the purpose of plenty if it turns into hunger? What is the value of beauty if it is taken away by time and decay?

Human life is always changing. What we were before, we are no longer; and what we are now, we will not always be. If nothing had an end, then its beginning would have had no purpose.

Lord, show me Your wisdom. Help me understand what I ask of You: How long will corruption continue? How much longer will human life go on as it is?

When will those who die be free from the wickedness of this world?

In Your mercy, bring to pass all that You have promised, so that those who think Your patience is weakness may see Your power.

Let those who do not understand—who see what has happened to us and to our city—realize that all these events fit within Your plan. You have called us Your people for the sake of Your great name.

But now, everything is ruled by death. I beg You, Lord, hold back the power of death. Let Your glory shine through. Let the beauty of

Your presence be revealed for all to see.

Close the gates of the grave so that it no longer takes in the dead.

Let the souls that have been waiting be released from their resting place. Many years have passed since the days of Abraham, Isaac, and Jacob—those who sleep in the earth, for whose sake You said You created the world.

Now, Lord, let Your glory be revealed without delay. Fulfill Your promises. Do not wait any longer."

When I finished praying, all my strength left me, and I felt completely weak.

Chapter 22~30

Then, the heavens opened before me, and I saw an incredible sight. A surge of strength filled me, reviving my spirit. I stood in awe as a voice from above spoke to me:

"Baruch, Baruch, why are you so troubled? Why is your heart weighed down with concern?

Who starts a journey and does not try to reach the end? Who sails out to sea without hoping to reach the shore safely?

Who promises to give a gift but never follows through? Wouldn't that be the same as stealing?

Who plants seeds in the ground but never expects to harvest them when the time is right? Or who plants a young tree and expects it to bear fruit before it has fully grown?

Doesn't a mother risk her child's life if she gives birth too soon? And isn't a house unfinished if it is built without a roof?

Tell me, are these things not true?"

I replied, "Yes, Lord, they are true."

Then He said, "If you understand this, then why are you troubled by things that are not yet clear to you? Why does your heart worry over matters you cannot yet understand?

Do you not realize that just as you remember those who live now and those who have passed, I also remember those who are yet to come?

When Adam sinned and death was passed on to all his descendants, I already knew how many souls would be born into this world. For each one, I prepared a place for the living and a place for the dead.

No soul will live again until the number I have set is complete. My Spirit is the source of all life, and the realm of the dead exists to receive those who have passed away.

But I have chosen to reveal to you what lies ahead. The time of redemption is closer now than ever before.

The days are coming when the books will be opened. These books hold the record of every sin committed by those who have turned away from Me. At the same time, the storehouses will be opened, revealing the righteousness of all who have walked in faith and obedience since the beginning of time.

In that day, you and many others will see with your own eyes the endless patience and mercy of the Most High. My patience has lasted through every generation, extending to both the righteous and the sinners.

The time is near when everything will be revealed and brought to completion."

I replied, "But Lord, no one knows how many things have already passed away, nor how many are still to come. I understand what has happened to us, but I do not know what will happen to our enemies or when You will bring justice."

The Lord answered, "You will be kept safe until the appointed time. When the sign that I have determined appears on the earth, it will mark the beginning of the final days.

This sign will bring fear and suffering upon the world. People will experience such deep trouble and pain that they will say in their hearts, 'The Mighty One has abandoned the earth.' But when they lose all hope, that is when the time of awakening will begin."

I asked, "Lord, how long will this time of suffering last? Will it continue for many years?"

The Lord replied, "This time has been divided into twelve distinct parts, each with its own purpose:

1. In the first part, there will be chaos and unrest.
2. In the second, the powerful will be overthrown.
3. In the third, many will die."

In the fourth part, violence and war will spread among the people.

In the fifth part, there will be famine, and the rain will stop.

In the sixth part, great disasters and earthquakes will shake the earth.

In the seventh part, spirits will roam, and demons will attack.

In the eighth part, fire will fall from the sky.

In the ninth part, there will be widespread oppression and terrible crimes.

In the tenth part, injustice and immorality will increase everywhere.

In the eleventh part, chaos will rise as all these troubles mix together.

And in the twelfth part, everything will reach its peak, as all these disasters become one.

"These times will be connected, each event affecting the others. Some will hold back their full force, while others will grow stronger. Those who live during this time will not realize they are witnessing the end of days. But those who have wisdom will understand, for this time has been measured as two periods of seven weeks."

I answered, "It may be a blessing to live and witness these events, but it could also be dangerous, for the risk of failure will be great. But Lord, let me ask—will these disasters only happen in one place, or will the whole earth experience them?"

The Lord replied, "These events will affect the entire world, and all people will experience them. But during this time, I will protect those who live in this land. When everything has been fulfilled, the Anointed One will begin to be revealed.

At that time, Behemoth will come out of its hidden place, and Leviathan will rise from the depths of the sea. These two great creatures, which I created on the fifth day of creation, have been kept for this very moment. They will provide food for those who survive. The earth will produce an abundance of crops, yielding a harvest ten thousand times greater than before.

Each vine will grow a thousand branches, each branch will have a thousand clusters, each cluster will hold a thousand grapes, and each grape will produce a large measure of wine. Those who once suffered from hunger will rejoice, for they will see miracles every day.

Each morning, a sweet breeze will come from My presence, carrying the scent of the fruits of paradise. In the evening, the clouds will release a dew that brings healing to all. And during those days, the storehouse of manna will open again from heaven, providing food for

My people. These will be the ones who have reached the end of time.

After these events, when the Anointed One's time is fulfilled, He will return in glory. All those who have died, hoping for His coming, will rise again. The places where the souls of the righteous have been kept will be opened, and they will come forth together as one united people.

The first will rejoice, and the last will have no reason to be sad, for they will all see that the promised end has come.

But the souls of the wicked will see these things and be filled with fear. They will realize that their time of suffering has arrived, and their destruction is near."

Chapter 31~34

After these events, I gathered the people and said, "Bring together all the elders of Israel, for I have something important to tell you." They listened, and soon the elders and the people assembled in the Kidron Valley. Standing before them, I spoke with urgency:

"Listen to me, O Israel, and hear my words. Pay close attention, descendants of Jacob, for I will share wisdom with you. Do not forget Zion, and do not let the suffering of Jerusalem fade from your hearts. Remember her pain and mourn for her.

The time is coming when everything you see now will fade away. It will be as if it never existed, disappearing into nothingness.

But as for you, if you prepare your hearts and hold tightly to the law, it will protect you when the Mighty One shakes the world. A time of great trouble is near, and it will test everyone. After this short period, Zion will be shaken so that it can be rebuilt. However, know this—the new city will not last forever. It will once again be torn down and left in ruins for a time.

But do not lose hope, for after that, Zion will be restored in glory, and its beauty will last forever.

Do not let the hardships you face now break your spirit. Instead, prepare yourselves for what is coming, for the challenges ahead will be even greater than those you have already endured. When the Mighty One renews His creation, the struggles of humanity will be more difficult than anything before.

Therefore, stay strong and keep watch.

For now, I ask you for this one thing: do not look for me for the next few days. I need time alone, for I must speak with the Mighty One. When I return, I will share with you what He reveals to me."

After saying these words, I turned and began to walk away. As the people watched me leave, they cried out loudly in sorrow:

"Where are you going, Baruch? Are you leaving us now, like a father abandoning his children? Will you leave us as orphans in our time of need?

Didn't your companion, Jeremiah the prophet, entrust us to your care? Didn't he say, 'Look after the rest of our brothers while I go to Babylon to prepare those in exile'?

And now, if you leave us too, it would have been better for us to die than to see you go."

Hearing their cries, I turned back to them and said:

"I would never abandon you or turn away from you! My heart is with you, and I remain responsible for you. But I must go now—for your sake and for the sake of Zion—to the Holy of Holies. There, I will seek guidance from the Mighty One and ask for wisdom.

Do not fear, for when my task is done, I will return to you."

With that, I continued on my way, leaving their voices of sorrow behind me as I went to seek the presence of the Mighty One.

Chapter 35~43

I, Baruch, went to the holy place, sat among its ruins, and wept bitterly. I cried out, "Oh, if only my eyes were endless springs of water, and my tears never stopped flowing, so I could properly mourn for Zion and grieve for Jerusalem.

How can I find the right words to express my sorrow for this holy city? Here, where I now sit in ruins, the chief priests once stood, offering holy sacrifices to the Mighty One. The altar was covered with the most fragrant incense, and the air was filled with the sounds of worship.

But now, all our glory has turned to dust, and our deepest desires have been reduced to ashes beneath our feet."

After I spoke these words, overwhelmed by grief, I fell asleep among the ruins. As I slept, I had a vision in the night.

I saw a vast forest spread across a wide plain, surrounded by tall mountains and jagged rocks. The forest stretched far and wide, covering much of the land. As I watched, a vine appeared before the forest, and from beneath the vine, a peaceful fountain began to flow. The water from the fountain grew stronger as it reached the forest, turning into mighty waves.

The waves crashed into the trees, uprooting them and knocking down the mountains that surrounded them.

The forest was brought low, and the once-great mountains were leveled. The waves grew stronger and stronger until only one tree remained—a single cedar. When the waves finally reached this last tree, they knocked it down with such force that the entire forest was completely destroyed. Nothing remained of its former greatness, and no one could even recognize where it once stood.

Then, the fountain and the vine moved peacefully together to a place near where the cedar had fallen. Strangely, the cedar was brought before the fountain and the vine, as if it were about to be judged.

As I watched, the vine opened its mouth and spoke to the cedar:

"Are you not the last of the cedars from this wicked forest? You allowed evil to thrive for years, and because of you, unrighteousness continued to spread.

You did nothing good. Instead, you reached into lands that were not your own, showing no kindness even to what belonged to you. You trapped those far from you in your web of wickedness, and those near you were crushed under your power.

You believed you could never be moved, but now your time has come, and your judgment is here.

So now, cedar, join the rest of the forest that has already fallen before you. Turn to dust, and let your ashes mix with theirs. Rest in suffering and sorrow, for your final punishment has not yet come. When the appointed time arrives, you will return to face an even greater judgment."

After these words, I saw the cedar catch fire and burn until it was completely consumed. Meanwhile, the vine grew stronger, spreading across the land. The ground around the vine became a beautiful plain, filled with flowers that never faded. The place radiated beauty and peace.

Then I awoke and rose from where I had been sleeping.

Shaken by the vision, I prayed earnestly to the Mighty One:

"O Lord, my God, You guide those who seek understanding. You have made Your law the source of life, and Your wisdom leads us on the right path.

Please help me understand this vision. Show me its meaning, for You know that my soul has always followed Your law and that I have never turned away from Your wisdom since my earliest days."

The Lord answered me:

"Baruch, I will explain the vision you have seen. The great forest surrounded by rugged mountains represents the kingdoms that have ruled over the earth. These are the kingdoms that have risen against Zion.

Listen, for the days are coming when the first kingdom that destroyed Zion will itself be destroyed. It will be conquered by another kingdom that will rise after it. This second kingdom will hold power for a time but will also fall. Then, a third kingdom will rise and rule, but it too will come to an end."

After this, a fourth kingdom will rise, stronger and more evil than the ones before it. It will rule over many people and spread its power far and wide, like a huge forest covering the land. This kingdom will see itself as the greatest, standing tall like the mighty cedar trees of Lebanon. But within it, truth will be hidden, and those who do wrong will find safety there, like wild animals hiding in a dense forest. Yet, when the right time comes, my Chosen One will appear, and he will tear down this kingdom and bring its power to an end.

The tall cedar tree that remained standing after the forest was destroyed represents this final kingdom. The words spoken to it by the vine show the judgment it will face. The vine and the fountain represent my Chosen One and the power of truth, which will bring peace and healing to the land. But the cedar tree and everything it stands for will be completely destroyed. This is the meaning of your vision, Baruch.

The last ruler of this kingdom will be captured and tied up, while his army will be wiped out by the sword. He will be brought to Mount Zion, where my Chosen One will confront him, revealing all his evil deeds and judging him for the actions of his armies. Nothing will be hidden—every crime will be exposed. Then, he will be put to death. My Chosen One will protect the rest of my people, those who have taken shelter in the place I have chosen for them. His rule will last forever, continuing until the world of corruption has come to its final end and everything I have planned has been completed. This, Baruch, is the vision you have seen and its true meaning.

Then I asked, "Lord, who will live to see these things happen? Who will be worthy of those times? Please allow me to speak freely and share my thoughts. I see many of your people who have turned away from your ways, rejecting your law. But I also see others who have left behind their foolish ways and taken refuge under your protection, choosing to follow your will. What will happen to them, Lord? How will they be treated in the last days? Will they be judged fairly, each one receiving what they deserve?"

The Lord answered, "I will tell you the answer. The good things I have spoken of will come to those who have believed in me, but those who have rejected me will receive the opposite. As for those who have come closer to me and those who have turned away, here is the meaning:

Those who once followed my law but later abandoned it, mixing with nations that do not know me, will be remembered for their former greatness before they fell, like tall mountains. But those who once lived without knowing the truth but later found it and joined my people will be honored for their transformation. Their later years will be like mountains, standing tall because of their change and dedication.

Time will pass from one period to another, and each season will follow the one before it, receiving from the past and continuing forward. In the end, everything will be judged at the right time. Those who belong to corruption will be taken by it, and those destined for life will receive it. The earth will be commanded to give back what it has taken, bringing forth all who have been buried when the moment comes."

Then the Lord said, "Baruch, keep your heart focused on everything I have told you. Understand and hold onto these words, for they bring comfort that will last forever. You will soon leave this world behind. You will forget everything that is temporary and never again think of the things of this life.

So now, go and tell your people to prepare for what is coming. Then return here and fast for seven days. After that, I will come to you again and reveal more to you."

Chapter 44~47

Then I, Baruch, left the place where I had received the message from God and returned to my people. I called for my oldest son, my trusted friend Gedaliah, and seven of the elders of the community. When they had all gathered, I spoke to them, saying:

"My time to leave this world has come, just as it does for everyone who lives on the earth. But you must remain strong and stay faithful to the law. Guard it carefully, and make sure the rest of the people do not turn away from the commandments of the Mighty One. Remember that the Creator we serve is fair and just in all his judgments.

Look at what has happened to Zion and consider what became of Jerusalem. These events prove that the Mighty One's judgment is right and that his wisdom is beyond our understanding. If you remain in awe of him, follow his laws, and do not turn away from his teachings, then the time of suffering will pass, and a time of comfort will come. Zion will be restored, and you will experience peace again.

The things of this world do not last. They are small and meaningless compared to what is coming. Everything that can decay will disappear, and everything that is mortal will come to an end. The pain and evil of this time will be forgotten, and no one will remember its sorrows. Those who chase after wealth and power now are wasting their efforts, for all their achievements will one day turn to nothing.

Do not place your hope in the present world, but in the one that is coming. That world will last forever. A new age is near, one where those who enter will never again be touched by corruption. In this new world, there will be no mercy for those who choose wickedness, and those who live there will never face destruction. It will be a world of purity and lasting peace.

This promised world is reserved for those who have sought wisdom, treasured understanding, and remained kind and faithful. They have held onto the truth and lived by it. This new world will belong to them, but those who reject it will face judgment.

So I urge you, warn the people with all your strength. This is your responsibility. If you guide and teach them, you will bring life to their souls."

When I finished speaking, my son and the elders were deeply saddened. They said to me, "Has the Mighty One truly humbled us so much that he will take you from us so soon? Will we be left in darkness, with no one to lead the people? Who will teach us the law? Who will help us know the difference between right and wrong?"

I replied, "I cannot go against the will of the Mighty One. But know this: Israel will always have a wise leader, and the tribe of Jacob will never be without a teacher of the law. You must prepare your hearts to follow the law and listen to those who have wisdom and understanding in the ways of the Lord. Stay strong and do not wander from the right path. If you do these things, the good promises I have spoken will come true for you, and you will not fall into the suffering I have warned you about."

I did not tell them the full truth about my departure, not even to my son, because the Mighty One had commanded me to keep it secret.

After I dismissed them, I told them, "I am going to Hebron because the Mighty One has sent me there." Then I left and arrived at the place where the Lord had spoken to me. There, I sat and fasted for seven days, waiting for what was to come.

Chapter 48~52

After seven days, I lifted my voice in prayer to the Mighty One and said:

"O Lord, you control the future, and time moves according to your will. You command the seasons to come and go, and they obey. The passing years follow the path you have set for them, and nothing can resist your eternal plan.

You alone know how long each generation will last, and you do not reveal all your secrets to everyone. You measure the size of the fire, control the winds, and see the highest heavens and the deepest darkness.

You keep track of those who will pass away and hold them according to your plan. You have prepared an eternal home for those yet to be born. You remember the very beginning of creation, when you brought everything into existence, and you have not forgotten the end, which has not yet come.

At your command, fire becomes obedient, changing into spirits that serve your will. By your word, you create life from nothing, and by your endless power, you sustain what is yet to come. With your wisdom, you guide all of creation, and you give understanding to the heavenly bodies so they can fulfill their purpose.

Countless angels stand ready before you, waiting for your command. They follow your instructions perfectly, carrying out their tasks in harmony and peace.

Now, listen to me, your servant, and hear my request. We, your creation, are here for only a short time, and just as quickly as we arrived, we return to dust. But to you, Lord, one hour is like an entire age, and one day is like a whole generation. Your understanding of time is beyond us.

Please do not be angry with us, for we are weak and temporary. Do not judge us too harshly, for what can we offer that is worthy of you? By your grace, we are brought into this world, yet we do not leave it by our own choice, but by your will.

We did not ask to be born, nor did we make plans for the afterlife. We have no power to stand under your judgment. What strength do we have that we could endure your wrath?

Show us compassion, Lord, and hold us up with your mercy. Only through your kindness and grace can we stand before you, the Creator of all things.

Look upon those who humble themselves before you, Lord, and rescue those who sincerely seek you. Do not take away the hope of your people, and do not end your mercy and salvation too soon. This is the nation you have chosen, the people you have set apart, unlike any other in the world.

Now I speak openly, sharing the thoughts in my heart. We have placed our trust in you, holding tightly to your law and cherishing its wisdom. Your commandments remain with us, and we know that as long as we follow them, we will not stumble.

We are blessed because we have stayed separate from the ways of other nations. We are united as one people, guided by the single law you have given us. This law belongs to us alone, and the wisdom you have placed in our hearts strengthens and sustains us."

When I finished praying, I felt weak, drained from pouring out my heart. Then the Mighty One answered me and said:

"Baruch, I have heard your prayer, and every word you have spoken has reached me. But understand this: my judgment follows its own course, and my law demands justice. I will answer you with the very words of your prayer and respond to your concerns.

Know this: nothing is ruined unless it first chooses to turn away from goodness, rejecting my patience and kindness. This is why I told you before that you would be taken up, and this promise still stands. The time of suffering will surely come. It will arrive suddenly, burning fiercely and sweeping through the world like a raging storm.

During those days, people will think they are living in peace, unaware that my judgment is near. The wise will become rare, and those who truly understand will be even harder to find. Many who once had knowledge will remain silent, withdrawing from the world.

Rumors will spread everywhere—some true, some false. Strange visions will appear, and many empty promises will be spoken—some completely baseless, others designed to deceive. Honor will turn into shame, and strength will crumble into disgrace. Confidence will break like fragile glass, and beauty will lose all value.

At that time, people will be confused and ask each other, 'Where has wisdom gone? Where can we find understanding?' And as these questions fill their minds, jealousy will grow among those who once felt small, and anger will consume those who were once peaceful. Wrath will stir up many, driving them to violence, and countless people will raise armies and spill blood. But in the end, they will all perish together, caught in the very destruction they sought to bring upon others."

These are the words spoken by the Mighty One—a warning for those willing to listen and a glimpse into what's coming.

"When that time arrives, it will be clear for all to see. The way things have always been will change. In the past, people sinned and hurt one another. Everyone chased after their own desires, thinking only of themselves and ignoring the laws of the Mighty One. They let their pride lead them away from the truth, choosing temporary pleasures instead of what lasts forever.

At that time, people's thoughts will be tested like metal in fire. The Judge will come in glory, and He will not be late. Deep down, everyone has known when they were doing wrong, but their pride kept them from admitting it. Because they ignored my law, they will mourn—not just for themselves, but for those still alive, because they will finally understand the weight of their mistakes and the price they must pay.

Then I, Baruch, spoke, saying, 'Oh Adam, what have you done to all your children? And Eve, what trouble has come from listening to the serpent? Because of you, so many have fallen into corruption, and countless souls will now face judgment.

But Lord, You know everything. You made Adam from the earth, and from him, all of humanity came to be. Only You know the number of people who have ever lived. You have seen how they turned against You, refusing to acknowledge You as their Creator. Because of this, they will face shame, and the law they ignored will stand against them when judgment comes.

But let's not speak of the wicked any longer—let's focus on the righteous. They have suffered much in this short life, but they will be rewarded with endless light in a world that never fades.

Now, Mighty One, tell me—what will happen to the living on that day? Will they keep these weak, mortal bodies, or will You change them along with the rest of creation?'

And He answered me:

'Baruch, listen carefully. Write down what I tell you because it is important. When the time comes, the earth will give back the dead. Right now, it holds them as they were when they died. And just as I placed them in the earth, they will return the same way. They will rise exactly as they were before.

This must happen so the living can see the truth—that the dead have come back, that those who were gone are now among them again. When people recognize their loved ones, they will know for certain that the resurrection is real, and this will make my judgment even more powerful. What has been promised will happen without a doubt.

After this, once that great day is over, both the wicked and the righteous will be changed. Those who lived in sin will suffer even more. Their appearance will become twisted, showing the pain and regret they feel. Their faces will be marked with fear and sorrow.

But the righteous—those who followed my law and sought wisdom—their beauty will shine. Their faces will glow with light, and they will be transformed in splendor. They will be ready to receive the everlasting world I have prepared for them. Their transformation will show their understanding and the wisdom they carried in their hearts.

Meanwhile, those who rejected my law and refused to listen to wisdom will be filled with sorrow. They ignored the truth, and now they will watch as those they once looked down on are lifted up in glory. The righteous will be raised to a place of honor, while the wicked will become unrecognizable, their suffering visible for all to see.

But the righteous will witness wonders beyond anything they ever imagined. They will see a world they never knew existed, a time and place hidden from them until now. Time will no longer have power over them. They will dwell in a new creation, as bright as the stars, as glorious as the heavenly messengers. Their beauty will change from one form to another, glowing brighter and brighter.

They will enter the Garden of Delight, and they will see the great living beings beneath my throne. They will behold the heavenly messengers who have been waiting for this moment.

Their glory will surpass even that of the messengers. The first ones will welcome the last, and the last will finally see those they had only heard about. Together, they will rejoice, for they have escaped the suffering of this world and are free from pain forever.

In this new world, they will never again feel sorrow. Their light will never fade, their joy will never end, and the hope they held onto will be fulfilled beyond their wildest dreams. Those who sought wisdom and truth will receive more than they ever expected. They will inherit the everlasting glory that was always meant for them.'

Then He said:

'Why, then, do people waste their lives chasing things that don't last? Why do they trade their souls for what has no true value? Long ago, they made their choice. They clung to a world that only brings suffering and grief, ignoring the promise of a place that will never decay. They turned their backs on the eternal glory I offered them, choosing a path that leads only to regret.'

And I responded:

'How can we forget those who are destined for suffering? Why do we mourn for those who have already died when a much greater sorrow is coming? Shouldn't we save our tears for the day of destruction ahead?

Still, I must ask—what will become of the righteous? How should they face these times? They should rejoice, for their suffering is only shaping them for what is to come. Why focus on the downfall of their enemies when they have a far greater reward ahead? They should prepare themselves for the blessings waiting for them. Their reward, stored up in the heavens, will be far greater than any of the hardships they endured in this world.'

After I had spoken, I became very tired, and I lay down and fell asleep in that place."

Chapter 53~54

I had a vision—an incredible and overwhelming sight that filled me with awe. I saw a massive cloud rising from the sea, unlike anything I had ever seen before. The cloud held swirling waters, both dark and light, mixing together in a chaotic motion. Within these waters, flashes of different colors appeared, blending in a way that was both beautiful and mysterious. At the very top of the cloud, powerful lightning shone brightly, cutting through the sky with an intense glow.

The cloud moved swiftly across the earth, as if driven by a purpose only it knew. Its shadow stretched far and wide, covering the land beneath it. Then, something strange happened—the cloud began pouring out its waters onto the earth. As I watched closely, I noticed that these waters were not the same.

First, the waters were dark, heavy, and full of gloom, spreading shadows wherever they fell. This lasted for a while, but then the waters changed. They became bright, filled with light and clarity, though they were not as strong as the dark waters. But after this brief moment of brightness, the darkness returned, heavier than before. This cycle continued twelve times—dark waters followed by bright, with the darkness always lasting longer and seeming more powerful.

At the end, the cloud released its final storm. This time, the dark waters were even blacker, and fire mixed within them. When these fiery waters hit the earth, destruction followed. Entire regions were consumed, the land trembled, and ruin spread everywhere.

But then, something incredible happened. The lightning that had been at the top of the cloud suddenly struck the earth with great force. It brought a light so brilliant that it illuminated the entire world. Under this glow, the places ruined by the dark waters began to heal. The destruction was undone, and life returned. But this lightning did more than just restore—it took control of the entire earth, bringing peace and shining with unstoppable brightness.

As I continued watching, I saw twelve rivers rise from the sea. These rivers moved toward the lightning, surrounding it and submitting to its power. Their waters were clear and pure, flowing in harmony with the light. A deep sense of awe and fear filled me, and I woke up from my vision, shaking.

I turned to the Mighty One, pouring out my heart before Him:

"Lord, You alone have known the secrets of the world since the beginning. Nothing is hidden from You. You see the deepest parts of the earth and the highest places in the heavens. You control time itself, setting each moment in its place.

Nothing is too hard for You, for all things obey Your word. With a single command, You hold together what is above and below. Your will shapes the course of history, and the ages move according to Your plan.

To those who honor You, You reveal what awaits them, giving them strength through Your wisdom. To those who do not know You, You break through their ignorance and open their eyes. You share Your secrets with those who trust in You, those who follow Your path with faith."

"Now, Lord, since You have given me this vision, I ask that You also help me understand its meaning. You have opened my mind, and I know I have received answers to my questions. You have shown me how to praise You, how to lift up my voice in honor of Your name. Yet, I realize that no matter how much I try, I can never fully express Your greatness.

Even if every part of my body could speak and every strand of my hair became a voice, it still would not be enough to give You the praise You deserve. My words could never fully describe the wonder of Your works. My heart cannot truly grasp the depth of Your beauty, which goes beyond all human understanding.

Who am I, that You would choose to show me these things? Compared to so many others, I am nothing—yet You have shared these mysteries with me. I am humbled by Your kindness. Blessed is the woman who gave birth to me, for she brought into the world someone who has been allowed to hear Your word. I will not stay silent. I will proclaim Your greatness and speak of Your wonders without end.

For who else, O Mighty One, can do what You do? Who else can understand the depth of Your wisdom? Every thought You have brings life, and all creation moves by Your eternal plan. You have placed the fountains of light beside You, and beneath Your throne, You have stored up treasures of wisdom, waiting for those who seek You.

But those who reject Your truth will face destruction. Judgment is waiting for those who refuse to follow Your ways. Adam sinned, and death came upon the world, but each person is responsible for their own choices. Every human, descended from Adam, chooses their own future—some preparing for eternal glory, others walking toward their own downfall. Those who believe in You will receive the reward of their faith.

But those who persist in wickedness, those who live in rebellion against Your truth, will soon face disaster. Their refusal to see Your wisdom will bring their own downfall. They have ignored the beauty of Your creation, which is a constant reminder of Your power. Their mistake is not just Adam's sin—it is their own choice. Each person is like their own Adam, shaping their fate through their actions.

Yet, Lord, I ask You again—help me understand the meaning of what You have shown me. Open my mind to the answers I seek. In the end, You will hold the wicked accountable for their deeds, and You will honor the faithful according to their devotion.

You rule over Your people with wisdom and justice, guiding them with love. But those who refuse to turn from evil will be removed from among the righteous. Lord, help me understand Your ways, for Your judgments are always right, and Your reign lasts forever."

Chapter 55~68

After I finished my prayer, I looked for a quiet place to rest. I sat beneath the wide branches of a tree, hoping for a moment of peace. The shade gave me relief, but my thoughts remained heavy. As I sat there, I thought about how much goodness sinners had turned away from. They had rejected the kindness and mercy freely given to them, choosing instead a path that would lead to their destruction. I wondered how they could ignore the suffering that awaited them, even though they knew the consequences of their choices. They refused to listen to the warnings, acting as if nothing would happen to them.

While I was deep in thought, trying to understand why people would reject such mercy, I suddenly sensed the presence of someone near me. I looked up and saw an angel standing before me. It was Remiel, the angel in charge of true visions, sent to me with a message. His presence was both powerful and gentle, and when he spoke, his voice was clear and strong.

"Why are you so troubled, Baruch?" he asked. "Why do these thoughts fill your mind with worry? If just hearing about the coming judgment makes you feel this overwhelmed, how will you handle it when you actually see it happen? If thinking about the day of the Mighty One weighs you down, how will you endure when it arrives?

"If simply hearing about the suffering of the wicked upsets you, how much more will you be shaken when you witness it with your own eyes? If even the mention of what will happen on that day fills you with sorrow, imagine how much greater your reaction will be when the power of the Almighty is revealed—bringing justice to some and joy to others.

"But because you have sincerely sought understanding, the Most High has sent me to explain your vision. Listen carefully, for the Mighty One has shown you the course of time—both past and future—from the moment the world was created until its final fulfillment. He has revealed the times when falsehood ruled and the times when truth prevailed.

"The great cloud you saw rising from the sea and spreading over the earth represents the entire span of history, from the beginning until the end. When the Mighty One decided to create this world, it started as something small, designed with great wisdom. By His command, it came into existence, structured according to His plan and guided by His perfect knowledge.

"The first dark waters that poured from the cloud represent the sin of Adam, the first man. When Adam disobeyed, death entered the world before its time. Grief and suffering became part of life, and pain took root in the hearts of humanity. Work became difficult, hardship increased, and pride led many away from goodness. The underworld, once still, became a force that constantly demanded more, taking life after life.

"Because of Adam's fall, people began to have children, but this came with sorrow and struggle. Humanity, once noble and pure, was humbled. The goodness that once filled the earth faded away like a passing mist. What could be darker than this? What could be more painful? This, Baruch, is the meaning of the first dark waters you saw in your vision—it represents the suffering that began with Adam's sin and has continued ever since."

"But from this darkness, even greater darkness was born. The suffering of the world grew worse because people turned away from their true purpose. They not only brought harm upon themselves but even disturbed the order of heaven. In the beginning, the angels were free from sin, but some of them chose to leave their place. They came down to earth and had children with human women, breaking the law of the Almighty. Because of this, they were punished—chained and cast into torment, a reminder of what happens when one disobeys the Creator.

"However, the majority of the heavenly messengers remained faithful. They did not give in to temptation but stayed true to their purpose. Meanwhile, the people on earth who had abandoned righteousness were destroyed in the great flood. These events marked the time represented by the first dark waters in your vision."

After those dark times, you saw bright waters. These waters represent the time of Abraham, the father of faith, and his descendants—his son, grandson, and those who followed in their ways. During this time, even though the law had not yet been written, people still followed it in their hearts. They lived by the commandments and believed strongly in the coming judgment. This was when hope for a renewed world began to grow, and the promise of a future life was planted. These bright waters symbolize the faith and goodness that shone during those days.

Then came the third waters, dark once more. These waters represent the sins that spread among the nations after the passing of righteous men. Evil flourished, especially in Egypt, where people acted cruelly and enslaved the children of Israel. But even this time of darkness did not last forever. The wickedness was judged, and justice was done.

Next, you saw the fourth waters, bright and shining. These waters represent the time of Moses, Aaron, Miriam, Joshua, Caleb, and others who stood with them. During this time, the eternal law was given, shining like a lamp for those in darkness. It brought hope to the faithful and warned the wicked of judgment. The heavens shook as the Mighty One revealed great truths to Moses. He showed him the principles of the law, the future of the world, and the design of the holy place, which reflected the heavens.

God also revealed to Moses the depths of the abyss, the weight of the winds, and the number of raindrops. He showed him how divine anger could be held back and how patience could be abundant. Moses was given wisdom, knowledge, and understanding. He saw the greatness of the Garden, the final judgment, and the future world. He learned about offerings, the coming ages, and places of justice and hope. These bright waters symbolize a time of revelation and divine guidance.

Then came the fifth waters, dark and ominous. These represent the sins of the Amorites, their evil magic, and the corruption they spread. Even the children of Israel fell into sin during the time of the judges, despite seeing many signs from their Creator.

The sixth waters, bright and clear, show the time of David and Solomon. During this time, Zion was built, the holy place was dedicated, and many sinful nations were defeated. The people made countless offerings, and the land was filled with peace. Wisdom was shared in gatherings, and understanding flourished. The holy festivals were celebrated with joy, and rulers judged fairly. People followed the commandments truthfully, and the land was full of mercy. Zion became a powerful and glorious center. These bright waters reflect the harmony and goodness of that time.

The seventh waters, dark and heavy, represent the deep corruption caused by Jeroboam, who led Israel into idol worship by setting up two golden calves. Many kings after him continued in wickedness. Jezebel's influence spread evil, and the people of Israel worshiped false gods. Because of this, the land suffered—rain stopped, and famine became so severe that people even ate their own children. Eventually, nine and a half tribes were exiled as punishment for their sins. The king of Assyria, Shalmaneser, took them away and scattered them in foreign lands. Meanwhile, the surrounding nations continued in their own wickedness. These are the dark waters of the seventh vision.

The eighth waters, bright and full of light, represent the faithfulness of Hezekiah, king of Judah. During his time, the Assyrian king Sennacherib gathered a huge army to destroy Judah, capture Zion, and wipe out the remaining two and a half tribes. But Hezekiah trusted in the Mighty One, praying for help. He asked God to see how Sennacherib planned to destroy them and lift himself up in pride. Because Hezekiah was faithful, God heard his prayer and acted.

God sent his messenger, Remiel, who destroyed Sennacherib's massive army. Among them, 185,000 leaders—each with their own troops—were burned up by fire from within, yet their clothes and weapons remained untouched, showing the power of the Mighty One. Because of this miracle, Zion was saved, Jerusalem was rescued, and the people of Israel were freed from their troubles. They rejoiced, spreading the name of the Mighty One everywhere. These are the bright waters you saw.

The ninth waters, dark and filled with evil, represent the terrible sins of Manasseh, Hezekiah's son. He committed terrible crimes—killing good people, corrupting justice, shedding innocent blood, and dishonoring marriages. He destroyed the altars, stopped the sacred offerings, and removed the priests from the holy place. He even built an idol with five faces, as if to challenge God himself.

Because of his actions, God's anger was unleashed. Zion was torn down, and judgment fell on the remaining two and a half tribes, sending them into exile. Manasseh's sins were so great that God's glory left the holy place. Because of his wickedness, Manasseh was given the title "the impious one," and his fate was sealed in fire. Even though he prayed for forgiveness in the end, his life remained deeply flawed. His final judgment was symbolized when a bronze horse statue he had fallen into melted, showing his punishment. These are the ninth dark waters you saw.

The tenth waters, bright and pure, represent the faith and devotion of King Josiah of Judah. He fully committed himself to the Mighty One with all his heart and soul. He removed idols, purified the sacred objects, and restored the altar offerings. He honored the righteous and reinstated the priests. He got rid of magicians, sorcerers, and false prophets, burning them and their remains.

Josiah brought back holy festivals and the sabbaths, cleansing the land of wickedness. He made sure no one in his kingdom acted unjustly. Because of his faithfulness, he will receive an eternal reward and be honored by God in the last days. For his sake, and for those like him, the promised blessings were prepared. These are the bright waters you saw.

The eleventh waters, dark and sorrowful, represent the disaster that has fallen upon Zion. The city, once full of God's glory, has been destroyed. The heavens do not celebrate its fall, though the nations mock and rejoice, saying that Zion, which once ruled over others, has now been conquered.

Do you think the Mighty One is pleased by this? Does He take joy in seeing His city ruined? No, He does not. But His justice will be carried out in the end, for He is always fair.

Because of these events, those scattered among the nations will face great hardships. They will live in shame, and as long as Zion remains in ruins, the wicked nations will thrive. Where Zion's holy incense once rose, now there is only the smoke of sin. The land once ruled by Zion is now filled with corruption.

The king of Babylon, who destroyed Zion, will become proud and believe he has won against the people of the Mighty One. But his pride will not last forever. In the end, God will humble him, and he will fall. These are the dark waters of the eleventh vision.

The twelfth waters, bright and full of hope, represent a time of renewal. After many hardships, your people will face their greatest struggle yet—so great that it will seem like they are about to be wiped out. But they will be saved, and their enemies will fall before them. It will be a time of great celebration and victory.

After this, Zion will be rebuilt. Offerings will once again be placed on its altars, and priests will return to their sacred duties. The nations will come to honor Zion and recognize the power of its God, though their respect will not be as deep as before. But even after this restoration, many nations will fall, their power crushed by divine justice. These are the bright waters you saw, pointing to Zion's future redemption and renewal.

Chapter 69~76

The last waters you saw were darker and blacker than all the ones before them. These waters came after the twelfth and were gathered together. They represent the fate of the whole world, not just one nation or group of people, but all of humanity. From the very beginning, the Most High separated the times, because only He knows the future. His knowledge covers all of history and even what has not yet happened.

He saw the evil that people would do in the future and identified six different kinds of wickedness. In the same way, He also saw six types of righteousness—the good things that faithful people would do in His sight, as well as the acts of kindness and salvation that He Himself would bring at the end of time. These last waters are different from the others. They are not just black mixed with black or bright mixed with bright. Instead, they represent the final moments—the end of this age.

Now, listen to what these last black waters mean. They show the days that are coming soon, when this world reaches its end, and everything people have done—both good and bad—has come to completion. In those days, the Mighty One will shake everything. People's hearts will be full of fear, and confusion will spread everywhere, even among kings and leaders.

People will begin to hate one another, turning against even their closest friends and neighbors. They will stir up anger, creating conflicts and division. Those who were once looked down upon will rise to power, while the honorable will be brought low. The weak will gain control over the strong, the poor will rule over the rich, and the wicked will claim victory over the brave.

During this time, wisdom will be silenced, and foolish people will take over. People's dreams and ambitions will fail, and their efforts will lead nowhere. Even the strongest leaders will struggle to hold things together. Everything that once seemed stable will fall apart, and fear will take hold of everyone's hearts and minds.

When this happens, the world will be thrown into chaos and suffering. Many will die in wars, caught up in violence and destruction. Others will not survive the hardships of their time. Some will even be betrayed by their own family members and fellow countrymen. These are the final black waters, marking the end of this age, the last chapter before everything changes. It is the deep darkness before the new beginning that the Mighty One has planned for the righteous.

Then, the Most High will show a clear and powerful sign to the nations and people whom He has already chosen and prepared. These nations will rise up and fight against the rulers still in power. This battle will be massive, shaking the earth itself. But even those who survive the war will not be safe.

If someone escapes the battle, they will die in a great earthquake. If they survive the earthquake, they will be destroyed by fire. If they somehow make it through the fire, they will starve to death. There will be no place for the wicked to hide—no part of the world will be free from disaster.

In the end, those who survive—whether they won their battles or were completely defeated—will be gathered together and handed over to my Servant, the Anointed One. At that time, the whole world will turn against its own people, bringing judgment upon them. No one will escape what has been decided, for the Most High will restore justice and righteousness.

However, the holy land, chosen and set apart from the very beginning, will show mercy to its own people. It will protect them and keep them safe from the overwhelming destruction of those days. This is the vision you have seen, Baruch, and this is its meaning. I was sent to reveal these things to you because your prayer reached the Most

High, and He has chosen to show you these hidden truths.

Now listen carefully about the bright lightning that comes after the final and darkest waters. This is the ultimate message: After all the signs I have told you about have happened—when confusion spreads across the nations, and the time of my Anointed One arrives—he will gather all the nations before him. Some will receive mercy and be saved, while others will face judgment and destruction.

Here is what will happen to the nations that are spared: Any nation that has not harmed Israel or mistreated the descendants of Jacob will be shown kindness. They will be allowed to continue, because even among these nations, some people will humbly submit to your people. But the nations that ruled over Israel or knowingly caused them harm will face judgment. They will not escape the consequences of what they have done.

After my Anointed One humbles and defeats all the powerful rulers of the earth, he will establish his throne and bring eternal peace. Joy will fill the world, and all of creation will finally find true rest. Healing will cover the earth like gentle rain, and sickness will disappear completely. There will be no more fear, sorrow, or suffering. The world will be wrapped in happiness like a warm cloak. No one will die too soon, and sudden disasters will be gone. Fighting, judgment, anger, envy, hatred, and all the things that bring pain and trouble will be removed forever. These are the things that have made life hard and bitter, but they will be no more.

Even the animals will live peacefully with people. Wild creatures will no longer be a danger, and they will live in harmony with humans. Snakes and dragons will come out of their hiding places and obey even a child. Women will no longer suffer pain in childbirth, and their children will be born easily, without struggle.

In those days, workers will no longer grow tired, and builders will not feel exhausted from their labor. The earth will produce food quickly and abundantly, and those who work the land will do so in peace. This is the time when everything that is temporary and corrupt will pass away, and something new and eternal will begin. It will be a world free from the evils of the past, full of goodness and lasting peace.

This, Baruch, is the meaning of the bright lightning that follows the final dark waters. It represents the completion of all things, the return of righteousness, and the beginning of a new world.

Then I responded and said:

"Who in all creation can compare to your endless goodness, O Lord? It is beyond human understanding.

Who can measure the depth of your mercy, which has no limits and lasts forever?

Who has the wisdom to grasp the fullness of your knowledge, which surpasses all human thought?

Who can describe the thoughts of your spirit, which move beyond what words can express?

Who among those born in this fleeting world can ever hope to reach such wisdom, unless you grant it through your mercy and compassion?

If your kindness did not extend to humanity—those you hold in your powerful right hand—we would have no strength to understand the mysteries of your ways. Only those you have chosen and written in your eternal book can be called to such knowledge.

But for those of us living now, if we recognize why we are here and submit ourselves to the One who led our ancestors out of Egypt with great power, we will reflect on the past and find joy in what has happened. We will look back with gratitude, seeing the purpose behind it all.

However, if we ignore the reason for our existence and refuse to acknowledge the One who freed us from slavery in Egypt, we will struggle to understand the meaning of the present. And when the truth finally becomes clear to us, we will be filled with sorrow, realizing too late that we missed the chance to live according to your purpose."

Then the Most High answered me and said:

"Now that I have shown you this vision and its meaning, listen carefully so you can understand what will happen to you after these events.

You will leave this world, but not through death. Instead, you will be set apart and kept safe until the end of time, when everything I have planned is fulfilled.

Now, go up to the top of the mountain, and I will show you the land stretched out before you. You will see not only the size of this world but also the image of all the lands where people live. You will see the high peaks of the mountains, the wide valleys, the deep seas, and the many rivers that flow across the earth. This will help you understand what you are leaving behind and the new destiny that awaits you.

But this will not happen immediately. You have forty days before this takes place.

During these days, go back to your people and teach them everything you can. Share wisdom and guidance with them, so they may turn their hearts toward the path of life. Teach them urgently, because the knowledge you give them will help them survive in the last days and avoid destruction. It will prepare them to live righteously in the time to come, when my glory is fully revealed."

Chapter 77

Then I, Baruch, left that place and gathered all the people together. I called everyone, from the most important to the least, so that they could hear my words. When they had all come, I spoke to them:

"Listen carefully, people of Israel, and think about the situation you are in. Look at how few of you remain from the twelve tribes. Once, we were a great nation, but now only a small group is left. The Lord gave His law to you and your ancestors, setting you apart from all the other nations. This law was your guide, a promise of righteousness, and the path to life.

But your brothers, the other tribes, disobeyed the commandments of the Most High, and He judged them fairly. He did not spare the first tribes that sinned, and He did not overlook the ones that came after. They were all taken away into exile, and none of them remained. Now, look at yourselves—you are still here with me because of His mercy.

So I urge you, follow the right path before the Lord. If you return to His commandments and live righteously, you will not suffer the same fate as your brothers. Instead, those who were scattered will come back to you. The Mighty One, whom you serve, is merciful and kind. The One you trust is full of grace and truth. He wants to bless you, not harm you.

Have you not seen what happened to Zion? Do you think the city itself committed a sin, and that is why it was destroyed? Do you believe the land itself did something wrong and was punished? No, you must understand this: Zion was innocent, but it was destroyed because of your sins and the sins of your brothers. Because some turned away from the right path, Zion, which had remained faithful, was handed over to its enemies."

When I finished speaking, the people all responded together:

"We remember, as much as we can, the great things the Mighty One has done for us and our ancestors. And for what we do not remember, we trust in His mercy, for He knows everything. But before you leave us, Baruch, we ask one thing: Write a letter of instruction and hope for our brothers in Babel. Strengthen them with your words, just as you have strengthened us. The leaders of Israel have been taken away from us, the lights that once guided us have gone out, and the wells from which we once drank have dried up. We feel lost, like wanderers in a forest or travelers in a desert."

I answered them:

"Do not lose hope, because the leaders, the lamps, and the fountains all come from the law. Even though I am leaving, the law will remain with you forever. If you focus on it and commit yourselves to its wisdom, you will always have light to guide your path, a leader to show you the way, and a source of life to sustain you.

But because you have asked me, I will write to your brothers in Babel to give them words of hope and guidance. I will also send a letter to the nine and a half tribes scattered in distant lands. One letter will be delivered by messengers, and the other I will send through a bird so that my words may reach them all."

So on the twenty-first day of the eighth month, I sat alone under the shade of an oak tree. In the quiet, I wrote two letters. One I gave to three men to take to those in Babel, and the other I gave to an eagle to carry to the nine and a half tribes living beyond the Euphrates River.

I spoke to the eagle, saying:

"You, noble bird, were created by the Most High to soar above all others. Now, I command you in His name to take this letter and deliver it quickly. Do not rest in a nest or perch on a tree until you have flown across the great waters of the Euphrates and reached the people to whom this message is sent.

Remember, it was a dove that brought back the olive branch to Noah, showing that peace had come after the flood. Ravens were sent to feed Elijah when he was in need. Even King Solomon, in his wisdom, sent birds to deliver his messages, and they obeyed him.

Now, do not hesitate or turn aside from your mission. Fly straight and fast, carrying out the command of the Mighty One, just as I have instructed you, so that His plan may be fulfilled through you."

With that, the eagle spread its wings and took flight, carrying the letter high above the earth. Meanwhile, I entrusted the other letter to the messengers headed for Babel. In this way, I made sure that the words of hope and instruction would reach all the scattered children of Israel.

Chapter 78~81

This is the letter that Baruch, the son of Neriah, wrote to the nine and a half tribes who were taken into exile and now live beyond the great Euphrates River. These are the words he sent to them:

"This is the message from Baruch, the son of Neriah, to my brothers who have been taken far from home. May grace and peace be with you from the Most High, who created us all and has never stopped loving us. From the very beginning, He has shown His love for us, even when we turned away from Him. He has never hated us, but like a caring father, He has disciplined us to lead us back to the right path.

I know this is true: even though we are scattered and living in different lands, we are still one people. We are all descendants of the twelve tribes of Israel, united by the covenant we made with our one true God. We share the same ancestors and, because of that, the same destiny.

Because of this, I have made sure to write this letter to you before my time comes to an end. I hope my words bring you comfort as you face the struggles that have come upon you. I also hope they remind you to grieve for the suffering of our brothers who are going through the same trials. Most of all, I want you to understand that our exile was not unfair or random—it was a just judgment. What has happened to us is less than what we actually deserved for turning away from the commandments of the Mighty One. But through this judgment, we have been given a chance to be worthy of the promises made to our ancestors for the future.

If you realize that the suffering we face now is meant to save us from being condemned in the end, then you will find a new kind of hope—one that lasts forever. But this hope requires action. You must turn away from the idols and false beliefs that led you away from the truth. If you do this, the Most High will remember you. He is the same God who made an eternal promise to our forefathers, vowing never to abandon or forget their descendants. Because of His great mercy, He will gather all of us who have been scattered and bring us back together.

Now, my brothers, think about what happened to Zion. Nebuchadnezzar, the king of Babel, came with his armies—not because Zion itself had sinned, but because we, its people, had turned away from the Most High. We failed to keep the commandments He gave us. Yet even in His judgment, He has been merciful, punishing us far less than we deserved.

When the enemy surrounded Jerusalem, messengers from the Most

High came to act. They tore down parts of the city's strong defenses and destroyed the iron fortifications that seemed impossible to break. But at the same time, they made sure to protect some of the sacred objects from the temple, hiding them so the enemy would not defile them. After that, they left the city's walls broken, the house of the Lord raided, and the temple burned, so that our enemies would not claim victory by their own strength.

Despite this destruction, our brothers were taken away to live in exile in Babel, far from the land of their ancestors. Those of us who remain are only a small group. This is the suffering I write to you about. I know how much comfort Zion once brought you. Just knowing that your brothers were safe in their land gave you peace, even though you had been taken far away.

But now, I want to share a message of hope. I, too, mourned for Zion. In my sorrow, I begged the Most High for mercy. I cried out to Him, asking if this suffering would last forever and if these troubles would ever come to an end.

The Mighty One, in His endless grace and kindness, heard my prayer. He sent me words of comfort to ease my pain and gave me visions to replace my despair with hope. In His mercy, He allowed me to understand the mysteries of time and revealed what will happen in the future.

So, my brothers, do not lose heart. The suffering we face now is not the end—it is only part of a greater plan. The Most High, whose wisdom and kindness go beyond what we can understand, has already prepared what comes next. Hold on to His law and the promises He gave our ancestors, because the days of restoration and glory are coming soon."

Chapter 82~84

My dear brothers, I am writing to you so that you may find comfort during these difficult times. Take courage, knowing that the Creator, who rules everything with justice, will bring judgment on our enemies. He will repay them for the wrongs they have done to us, for the evil they have spread, and for how they have ignored His laws. Also, remember that the time the Most High has planned is coming closer, and His mercy is reaching out to us. His final judgment is near, and when it comes, everything that has been broken will be restored.

Look around and see what is happening even now. The nations of the world seem to be thriving, yet they continue to do evil, piling sin upon sin. But their success will not last—it will disappear like smoke in the air. We see their power and arrogance as they refuse to honor the Mighty One, but their strength is as small as a drop of water, quickly gone. They may act as if they can resist the laws of the Most High, but in the end, they will be as insignificant as dust.

Think about the way they proudly display their wealth and power while ignoring the commandments of the Mighty One. Their riches and glory will fade away like mist, leaving nothing behind. Consider their beauty, which is covered in sin and wrongdoing—it will wither as quickly as dry grass in the scorching sun. Their cruelty and oppression may seem unstoppable, but like waves crashing on the shore, they will break and disappear.

And what about their pride? They lift themselves up, bragging about the power they have been given, yet they refuse to acknowledge the One who gave it to them. Their arrogance will vanish like a cloud blown away by the wind, leaving no trace behind.

The Most High will not wait forever. He will bring about His plan at the right time. He will judge everyone who lives on this earth. Every action, every sin, and even the hidden thoughts deep in people's hearts will be revealed. Nothing will remain secret—everything will come to light before all, and those who are guilty will be held responsible.

So do not let what you see now make you lose hope. Instead, hold on to faith and trust that the promises of the Most High will come true. Do not focus on the temporary success of the nations that seem to be thriving now. Instead, remember the inheritance that has been promised to us in the end. The pleasures of this world will disappear like shadows, but the reward waiting for us is eternal.

The end of this age is coming quickly, and everything in it will pass away. When this time is over, the great power of the Mighty One will be revealed for all to see, as everything is judged. On that day, His righteousness will be shown clearly, and the sins of the nations will no longer be hidden. Then, His everlasting kingdom will shine in full glory, unshaken and eternal. Let this be the hope that carries us through every trial.

So, my brothers, I am writing to encourage you to stay strong in your faith. Hold on tightly to what you have learned so that you do not lose everything—both in this life and in the next. Think about this: everything that exists now, everything that has already happened, and everything that will come in the future—none of it is completely good or completely bad. Life is always changing, reminding us that nothing stays the same forever.

Good health eventually turns into sickness. Strength fades into weakness. Power never lasts and will one day disappear. Even the energy of youth will one day give way to old age and, finally, to the end of life itself.

Beauty and grace, which people celebrate, will eventually fade, leaving only decay behind. The innocence of childhood, full of promise, often turns into disappointment and shame. Honor and glory, no matter how great, will one day be forgotten. Happiness, no matter how bright, will eventually turn into sadness.

Even the loudest voices of pride will be silenced, crumbling into dust. Riches and possessions, which seem so secure, cannot stop the reality of death. Desires that control and consume people will lead only to an end they cannot escape. Pleasure that seems enjoyable for a moment will bring judgment later.

Lies and falsehoods may seem to win for a time, but they will always be exposed in the end. Sweet words that once comforted will turn bitter when the truth is revealed. Friendships that are not built on honesty will eventually break, turning into betrayal.

Since we already see these things happening, do you really think they will go unnoticed? When everything reaches its end, every hidden truth will be brought into the light.

Now, listen carefully. While I am still with you, I am sharing this wisdom. More than anything, I urge you to follow the commandments of the Mighty One. Before I leave this world, I want to remind you of His instructions so that you can live by them.

Remember what Moses did long ago. He called on heaven and earth to witness his words when he warned the people, saying, "If you disobey the law, you will be scattered, but if you follow it, you will be blessed and grow strong." He spoke these words when all twelve tribes were still together in the wilderness. But after he died, you turned away from his teachings, and because of that, everything he warned about has now happened.

Now I speak to you after all your suffering. Moses warned you before these troubles came, and his words came true because you abandoned the law. In the same way, I now tell you this: If you listen and follow the teachings I give you, the Mighty One will bless you with all the good things He has prepared for you.

Let this letter be a witness between us. Let it remind you of the commandments of the Mighty One, and let it stand as my defense before Him. Never forget Zion, the law, the holy land, your brothers, the covenant, or your ancestors. Keep the festivals and the Sabbaths. Pass this letter down to your children, just as your ancestors passed down the law to you.

Always seek the Mighty One with all your heart and soul. Pray sincerely and ask for His mercy, so that He does not count all your sins against you. Instead, may He remember your honesty and your desire to follow Him. Teach your children to do the same, so they can walk in His ways and receive the blessings He has promised to those who remain faithful.

If our Creator does not judge us with mercy, then we are all truly hopeless. This life is short, and everything we love will one day fade away. Do not fool yourselves into thinking that your actions—whether good or bad—will go unnoticed. The Mighty One sees everything, even the hidden thoughts of the heart and the deepest secrets of the soul. Nothing is hidden from Him; everything will be revealed when the time of judgment comes.

Think about how short life is. No matter how strong a person is, sickness and weakness will eventually come. Youthful energy fades into old age. Power and influence do not last forever—they always disappear with time. Beauty, no matter how admired, will one day fade into ruin.

Many chase after wealth and status, but these things do not last. Pride turns into humiliation, and those who once had great honor will eventually be forgotten. Riches disappear, and the happiness people seek in them will eventually be replaced with sorrow. Even the loudest and most powerful voices will one day be silent, and no amount of wealth can keep someone from the grave.

People chase after their desires, but in the end, all of it leads to death. Those who live only for pleasure will one day face judgment. Lies and deceit may work for a while, but eventually, the truth will be revealed, and they will be condemned. Friendships built on dishonesty will crumble, leaving only betrayal and pain.

Do you really think that all these things will go unpunished? When everything comes to an end, the truth will be made clear, and every action will be judged. The Mighty One will bring His time of judgment quickly, and everything will happen just as He has planned. He will judge the world with perfect justice, and nothing will be hidden from Him. Every secret thought and every hidden deed will be brought into the light.

So do not put your hope in the temporary pleasures of this world. Instead, focus on what has been promised to us—the eternal inheritance that awaits those who remain faithful. When the end of this age comes, the great power of the Mighty One will be revealed. On the day of judgment, all creation will see His justice. Let us prepare our hearts now and live by His commandments, so that we may stand before Him blameless and receive His mercy and grace.

Chapter 85~87

In the past, our ancestors had righteous leaders—prophets and holy men who stood between us and the Mighty One. These faithful people guided us and prayed on our behalf when we failed. Because of their goodness and strong faith, the Mighty One heard their prayers and forgave our mistakes. Their prayers cleansed us and brought us back into His favor.

But now, things are different. Those righteous people have passed away, and the prophets who spoke the words of the Most High are no longer with us. We have been taken from our land, and Zion, our sacred home, has been taken from us. All that remains is the Mighty One and His holy law. Yet even in this time of loss, we still have hope, because His law is our guide and proof of His promise to us.

If we fully turn back to Him and live according to His commandments, He will not only return what we lost, but He will give us something far greater than we can imagine. What we lost was temporary—it could be destroyed or taken away. But what the Mighty One will give us in return will last forever. It is a promise that should strengthen us and keep our faith strong.

That is why I have also written to our brothers in Babel, so they too may understand these truths and find comfort. Keep these words in your hearts, because we are still under the mercy and freedom that the Most High has given us. He is patient and kind, showing us what is coming and helping us prepare for the future. This is a great gift, giving us the chance to be ready.

Before the day of judgment arrives and the truth is revealed to all, we must prepare ourselves. Let us hold on to hope instead of falling into despair. Let us seek eternal peace with our ancestors instead of facing suffering with those who rejected the ways of the Mighty One. The strength of this world is fading, and time is running out. The final moments are approaching, and the end of the journey is near.

So be ready, just as travelers prepare for the last part of their trip. When you leave this life, may you find peace and not punishment. The Mighty One will fulfill everything that has been foretold. When that time comes, there will be no more chances to repent, no more moments of mercy, and no more opportunities to pray for forgiveness.

Love offerings will cease, cries for help will go unanswered, and no prophets or righteous people will be left to speak for us. The time to act will be over, and only judgment will remain.

On that day, those who lived in corruption will face destruction, cast into fire and eternal judgment. The world follows one law, given by One Creator, and in the end, everyone will face the same decision. The Mighty One will separate the righteous from the wicked. Those who can be forgiven will be saved and purified, but those who refuse to change will be cast away.

When you receive this letter, read it carefully when you gather together, and reflect on its words, especially during times of fasting. Let it remind you of the covenant and encourage you to stay strong. Remember me, as I also remember you, and let this letter keep us connected in shared faith and hope.

This is the end of the letter of Baruch, son of Neriah.

After writing these words, I folded and sealed the letter. Then, I tied it securely to the neck of an eagle and released it into the sky, sending it to our brothers far away. And so, I entrusted these words to their journey, hoping they would bring guidance, comfort, and hope to all who received them.

Apocalypse of Abraham

Abraham was known for his kindness, fairness, and generosity. He lived near a place called Dria the Black, at a crossroads where many travelers passed through. He welcomed everyone—rich or poor, kings or commoners, strong or weak. No matter who they were, Abraham treated them with kindness because he was a good and just man who loved people.

One day, the Lord called the archangel Michael and said, "Go to my servant Abraham and remind him that his time on earth is coming to an end. I have blessed him greatly, making his descendants as countless as the stars in the sky and the sand on the shore. He has lived a life full of goodness and generosity. Now, his time has come."

Michael, who sat before the Lord, left heaven and went to find Abraham in Dria the Black. When he arrived, he saw Abraham working in the field with his servants and some young men. The archangel approached him and said, "Greetings, honored father, chosen one of the Lord, beloved friend of the King of Heaven."

Abraham replied, "Greetings to you, mighty one of God's army! You are more radiant than any man I have ever seen. Tell me, young man, where do you come from, and why do you shine so brightly?"

Michael answered, "Righteous Abraham, I come from the Great City. The Great King has sent me to His chosen friend to tell him to prepare himself, for the Lord is calling him."

Abraham nodded and said, "Very well. Let us go back to my home." Then he called his servants and said, "Go to the field and bring two of my horses. Prepare them so I may ride one, and my guest may ride the other."

But Michael replied, "Do not bring the horses. I do not ride animals with four legs. Let us walk together, righteous one."

As they walked, they passed by a tall and sturdy cypress tree. Suddenly, the tree cried out, "The Lord calls you, Abraham!" But Abraham remained silent, unsure if the angel had heard it.

When they reached Abraham's home, they sat down. Isaac, Abraham's son, saw the angel and said to his mother, Sarah, "Look at the man sitting with my father. He does not look like any ordinary person."

Isaac ran to the angel, bowed before him, and the angel blessed him, saying, "May God give you all the blessings He has given to your father and mother."

Abraham turned to Isaac and said, "Bring a basin and fill it with water so we can wash our guest's feet."

Isaac ran to the well, filled a basin, and brought it back. As Abraham washed the angel's feet, he sighed deeply and began to cry. Seeing his father weep, Isaac also started to cry, and their tears fell together. The angel, moved by their sadness, wept as well. As his tears fell into the basin, they turned into precious stones.

When Abraham saw this, he gathered the jewels and kept their meaning in his heart.

Then Abraham told his son, "Go prepare two beds carefully. Set candles in the candlesticks, lay out the table, light incense, and spread fragrant herbs on the floor so the room smells sweet. Light seven candles so that we may celebrate this guest, who is greater than any man and mightier than kings."

Isaac did everything as his father instructed.

Abraham and the angel went into the prepared room. They sat down on separate beds with a table of food between them. Then the angel returned to the Lord and said, "Lord, I have seen Abraham's righteousness, kindness, and incredible strength. I cannot bring myself to tell him about his approaching death because I have never met anyone like him on earth."

The Lord replied, "Go back to my friend Abraham. Eat the food he has prepared, and I will send My Spirit to his son Isaac. In a dream, I will reveal to him that his father's time is near. You will interpret the dream so that Abraham may understand that his time has come."

The archangel said, "Lord, heavenly beings do not eat or drink. How can I sit and eat with Abraham?"

The Lord replied, "Do not worry. I will send spirits to make the food disappear from your hands and mouth, as if you were eating. This will bring joy to Abraham and his family. Also, explain Isaac's dream so they understand what is about to happen."

The archangel returned to Abraham, and they ate together. As usual, Abraham said a prayer before the meal. After eating, they prayed again and then rested on their beds.

Isaac turned to his father and said, "I want to stay here and listen to our guest."

But Abraham replied, "No, my son. Go to bed and rest. We must not trouble our guest."

Isaac obeyed, received his father's blessing, and went to his room.

Later that night, Isaac had a dream that frightened him. He ran to his father's room, where Abraham was still with the archangel, and cried, "Father Abraham, please open the door! Let me hold you before they take you away from me!"

Abraham got up and opened the door. Isaac ran inside, embraced his father, and wept loudly. Abraham also wept, and when the archangel saw them, he wept too.

Abraham gently asked Isaac, "My dear son, tell me what you saw in your dream that has upset you so much."

Isaac replied, "I saw the sun and the moon resting on my head, shining brightly in all directions. At first, I was happy, but then the heavens opened, and a glowing man came down. He removed the sun from my head and took it to heaven. Then he did the same with the moon. I begged him, 'Please, do not take them away from me!' But he said, 'Let them go. The Lord of Heaven has called for them.' Although they left some of their light behind, I felt heartbroken."

Abraham sighed and said, "The sun you saw, and the glowing man from heaven, must mean that my time to leave has come." He then turned to the angel and said, "Oh, how amazing! But I fear you are the one who has come to take my soul from me."

The archangel replied, "I am the angel sent to bring you news of your passing. You will go to the Lord as promised in your covenant."

Abraham answered, "Now I understand that you are here to take my soul, but I will not go willingly!"

The angel returned to the Lord and reported everything that had happened, including Abraham's refusal, saying, "He will not surrender."

The Lord said to the archangel, "Go back to my friend Abraham and remind him: I am the Lord, his God, who led him to the Promised Land. I blessed him with descendants as countless as the sand on the shore and the stars in the sky. How dare he resist me? Does he not know that since the time of Adam and Eve, all people have died? Kings, ancestors, and all of humanity have faced death because no one is immortal.

"But I have not sent him sickness, suffering, or the grim reaper to take him away. Instead, I sent my archangel Michael with this message so Abraham could prepare himself. Why does he resist my messenger? Does he not know I could send the angel of death, whose presence he could not endure?"

The archangel returned to Abraham and repeated the Lord's words. Abraham wept and said, "Mighty angel of heaven, though I am a sinner, you have honored me. Please grant me one last request. The Lord has always answered my prayers and given me what I asked for. I know I cannot escape death, but before I die, let me see all the people of the earth and their deeds while I am still alive. After that, I will surrender myself completely."

The archangel returned to heaven and told the Lord about Abraham's request.

The Lord said, "Place my servant Abraham in the chariot of the cherubim and bring him up to heaven."

Then sixty angels prepared the chariot. Abraham was lifted up on the clouds. As he traveled, he saw another chariot behind him and groups of people below.

In one area, he saw people committing terrible sins and cried out, "Lord, let the earth open and swallow them!"

In another place, he saw people stealing and harming others and shouted, "Lord, send fire from heaven to destroy them!"

Fire came down and consumed them.

A voice from heaven commanded, "Take Abraham away from this sight so he will not see the people any longer. If he continues watching their sins, he will destroy them all. But I do not wish for anyone to perish. I want the wicked to repent and live. Take Abraham to the first gate of heaven so he may witness the final judgment and humble himself even more."

The archangel turned Abraham's chariot and brought him to the first gate of heaven. There, he saw two paths—one narrow and difficult, the other wide and easy.

On the narrow path, only a few souls were walking, each guided by an angel.

On the wide path, there were many souls, but they looked wounded and suffering, being led by different beings.

Then Abraham noticed a powerful figure sitting on a golden throne. Sometimes, the figure wept, pulling at his hair and beard when he saw the many souls on the wide path. Other times, he rejoiced when he saw the few souls walking the narrow path.

Abraham turned to the archangel and asked, "Who is this man who switches between sorrow and joy?"

The archangel answered, "This is Adam, the first man, created to bring beauty to the world. He rejoices when he sees souls on the narrow path because it leads to life. But when he sees so many souls on the wide path, which leads to destruction, he mourns deeply."

As they spoke, two angels arrived, bringing countless souls before Adam. Some were sent down the narrow path, while others were turned away.

Then Abraham saw another golden throne at a large gateway. It shined like fire, and a man sat on it, resembling the Son of God. In front of him was a massive table, and two angels stood beside him.

One angel held a set of scales, and the other held a scroll listing all the temptations and sins of humanity. The man judged each soul, deciding their fate.

The angel on the right recorded virtues, while the angel on the left noted sins. Some souls were condemned, others were set free, and a few were placed in the middle.

Abraham asked the archangel, "What is this I see before me?"

The angel replied, "These are the judges, and they pass judgment on every soul that comes before them."

Abraham watched as one soul was brought forward.

An angel said, "This soul has an equal number of good and bad deeds. Erase its record, for it will neither be saved nor condemned. Place it in the middle."

Abraham then asked, "Who are these judges and the glowing angels surrounding them?"

The archangel said, "Lord, heavenly beings do not eat or drink. How can I sit and eat with Abraham?"

The Lord replied, "Do not worry. I will send spirits to make the food disappear from your hands and mouth, as if you were eating. This will bring joy to Abraham and his family. Also, explain Isaac's dream so they understand what is about to happen."

The archangel returned to Abraham, and they ate together. As usual, Abraham said a prayer before the meal. After eating, they prayed again and then rested on their beds.

Isaac turned to his father and said, "I want to stay here and listen to our guest."

But Abraham replied, "No, my son. Go to bed and rest. We must not trouble our guest."

Isaac obeyed, received his father's blessing, and went to his room.

Later that night, Isaac had a dream that frightened him. He ran to his father's room, where Abraham was still with the archangel, and cried, "Father Abraham, please open the door! Let me hold you before they take you away from me!"

Abraham got up and opened the door. Isaac ran inside, embraced his father, and wept loudly. Abraham also wept, and when the archangel saw them, he wept too.

Abraham gently asked Isaac, "My dear son, tell me what you saw in your dream that has upset you so much."

Isaac replied, "I saw the sun and the moon resting on my head, shining brightly in all directions. At first, I was happy, but then the heavens opened, and a glowing man came down. He removed the sun from my head and took it to heaven. Then he did the same with the moon. I begged him, 'Please, do not take them away from me!' But he said, 'Let them go. The Lord of Heaven has called for them.' Although they left some of their light behind, I felt heartbroken."

Abraham sighed and said, "The sun you saw, and the glowing man from heaven, must mean that my time to leave has come." He then turned to the angel and said, "Oh, how amazing! But I fear you are the one who has come to take my soul from me."

The archangel replied, "I am the angel sent to bring you news of your passing. You will go to the Lord as promised in your covenant."

Abraham answered, "Now I understand that you are here to take my soul, but I will not go willingly!"

The angel returned to the Lord and reported everything that had happened, including Abraham's refusal, saying, "He will not surrender."

The Lord said to the archangel, "Go back to my friend Abraham and remind him: I am the Lord, his God, who led him to the Promised

Land. I blessed him with descendants as countless as the sand on the shore and the stars in the sky. How dare he resist me? Does he not know that since the time of Adam and Eve, all people have died? Kings, ancestors, and all of humanity have faced death because no one is immortal.

"But I have not sent him sickness, suffering, or the grim reaper to take him away. Instead, I sent my archangel Michael with this message so Abraham could prepare himself. Why does he resist my messenger? Does he not know I could send the angel of death, whose presence he could not endure?"

The archangel returned to Abraham and repeated the Lord's words. Abraham wept and said, "Mighty angel of heaven, though I am a sinner, you have honored me. Please grant me one last request. The Lord has always answered my prayers and given me what I asked for. I know I cannot escape death, but before I die, let me see all the people of the earth and their deeds while I am still alive. After that, I will surrender myself completely."

The archangel returned to heaven and told the Lord about Abraham's request.

The Lord said, "Place my servant Abraham in the chariot of the cherubim and bring him up to heaven."

Then sixty angels prepared the chariot. Abraham was lifted up on the clouds. As he traveled, he saw another chariot behind him and groups of people below.

In one area, he saw people committing terrible sins and cried out, "Lord, let the earth open and swallow them!"

In another place, he saw people stealing and harming others and shouted, "Lord, send fire from heaven to destroy them!"

Fire came down and consumed them.

A voice from heaven commanded, "Take Abraham away from this sight so he will not see the people any longer. If he continues watching their sins, he will destroy them all. But I do not wish for anyone to perish. I want the wicked to repent and live. Take Abraham to the first gate of heaven so he may witness the final judgment and humble himself even more."

The archangel turned Abraham's chariot and brought him to the first gate of heaven. There, he saw two paths—one narrow and difficult, the other wide and easy.

On the narrow path, only a few souls were walking, each guided by an angel.

On the wide path, there were many souls, but they looked wounded and suffering, being led by different beings.

Then Abraham noticed a powerful figure sitting on a golden throne. Sometimes, the figure wept, pulling at his hair and beard when he saw the many souls on the wide path. Other times, he rejoiced when he saw the few souls walking the narrow path.

Abraham turned to the archangel and asked, "Who is this man who switches between sorrow and joy?"

The archangel answered, "This is Adam, the first man, created to bring beauty to the world. He rejoices when he sees souls on the narrow path because it leads to life. But when he sees so many souls on the wide path, which leads to destruction, he mourns deeply."

As they spoke, two angels arrived, bringing countless souls before Adam. Some were sent down the narrow path, while others were turned away.

Then Abraham saw another golden throne at a large gateway. It shined like fire, and a man sat on it, resembling the Son of God. In front of him was a massive table, and two angels stood beside him.

One angel held a set of scales, and the other held a scroll listing all the temptations and sins of humanity. The man judged each soul, deciding their fate.

The angel on the right recorded virtues, while the angel on the left noted sins. Some souls were condemned, others were set free, and a few were placed in the middle.

Abraham asked the archangel, "What is this I see before me?"

The angel replied, "These are the judges, and they pass judgment on every soul that comes before them."

Abraham watched as one soul was brought forward.

An angel said, "This soul has an equal number of good and bad deeds. Erase its record, for it will neither be saved nor condemned. Place it in the middle."

Abraham then asked, "Who are these judges and the glowing angels surrounding them?"

The Vision of Ezra

Introduction

The Vision of Ezra is an ancient text that describes what happens after death, showing the rewards of the righteous and the punishments of the wicked. It is written as a series of visions experienced by Ezra, also known as Salathiel.

Although the text claims that Ezra wrote it, scholars believe it was actually written sometime between the 2nd and 10th centuries AD. The oldest copies that still exist today are in Latin and date back to the 11th century. However, experts think the original text was written in Greek, suggesting it came from a Hellenistic background. The exact time it was written is unknown, but its style and themes are similar to Christian writings from the 3rd and 4th centuries.

The Vision of Ezra has been found in seven Latin manuscripts from the 11th to the 13th centuries, with important copies stored in the Vatican Library and other monasteries in Austria. The similarities between these copies show that the text was well-known and influential in medieval Christian teachings.

The story follows Ezra on a deep spiritual journey. He asks for the strength to witness God's judgment on sinners, and in response, seven angels guide him through the different levels of punishment. Along the way, he sees:

- Fiery Gates – These are guarded by lions that breathe flames.
- Punishment of Sinners – The wicked suffer in terrifying ways, such as being attacked by wild dogs or burned in fire.

- Reward of the Righteous – Those who have lived well pass safely through fire, which purifies them and prepares them for salvation.

These powerful images highlight the difference between the fate of good and evil people, showing God's justice and mercy.

The Vision of Ezra reflects early Christian beliefs about the afterlife and final judgment. It describes heaven and hell in great detail, offering lessons on morality, repentance, and salvation. The text also emphasizes free will, showing that people's choices in life determine what happens to them after death.

This text has similarities to other apocalyptic writings, such as the Greek Apocalypse of Ezra and 2 Esdras (also called 4 Ezra). All of these focus on visions of the afterlife and divine judgment, but the Vision of Ezra is shorter and tells a more focused story. Its unique feature is Ezra's journey through different spiritual realms, guided by angels.

The Vision of Ezra gives a strong picture of early Christian beliefs about what happens after death, God's judgment, and the choices people make in life. Its detailed descriptions and important lessons still have meaning today, offering a glimpse into the religious and spiritual ideas of the past.

The Book of Ezra the Scribe, Who Is Called Salathiel

Vision I

Introduction (III. I 3).

In the 30th year after our city was destroyed, I, Salathiel, also known as Ezra, was in Babylon. I lay on my bed, feeling troubled, as many thoughts filled my mind. I saw how Zion had been left in ruins while

Babylon's homes were filled with riches. My heart was overwhelmed, and I cried out to God in fear.

I said, "Lord, didn't you speak from the beginning when you created the earth? You alone shaped everything, commanding the dust to form Adam. You breathed life into him, and he became a living being. You placed him in the paradise you had prepared before the earth existed.

You gave him one command, but he disobeyed it. Because of this, you sentenced him and his descendants to death. From him came many nations, tribes, and languages—too many to count. They all followed their own ways, committing evil, but you did not stop them.

Then, in time, you sent a great flood to destroy the people of the world because of their wickedness. They all perished, just as Adam faced death. But you spared one man and his family, and from them, all the righteous people were born.

As people multiplied again, their wickedness grew worse than before. When they turned away from you, you chose one man, Abraham. You loved him and revealed your plans for the future. You made an everlasting covenant with him, promising to never abandon his descendants.

You gave him Isaac, and to Isaac, you gave Jacob and Esau. You chose Jacob's descendants as your own people, while you rejected Esau. Jacob's family grew into a great nation. When you brought them out of Egypt, you renewed your covenant with them and led them to Mount Sinai.

You shook the heavens and the earth, causing the sea to tremble and the world to fear your power. Your glory passed through fire, earthquakes, wind, and cold to give Jacob's descendants your Law and commandments. But even then, you did not remove their sinful hearts.

Because of Adam, sin had taken root, and everyone born after him inherited his weakness. Even with your Law, people still chose evil.

As time passed, you raised up a servant, David, and commanded him to build a city for your name, a place where offerings would be made to you. For many years, this was done. But the people of that city also sinned against you, just as Adam and his descendants had done. They followed their sinful hearts, so you allowed their enemies to take the city.

Seeing this, I wondered in my heart: Are the people of Babylon really better? Have you abandoned Zion for them? Since I arrived here, I have witnessed endless sins. For 30 years, I have seen terrible wrongdoing. My heart is troubled because I see you allow sinners to thrive, while your own people have been destroyed, and your enemies remain untouched.

You have not made it clear how anyone can understand your ways. Has Babylon really done better than Zion? Is there any other nation you favor more than Israel? What other people have kept your covenant like Jacob's descendants? Yet, they have not been rewarded, and their hard work has not paid off.

I have traveled among many nations and seen that they are successful, even though they do not follow your commands. If you weigh our sins against those of the rest of the world, the balance would not tip much in either direction. Have the people of the world ever stopped sinning before you? Has any nation truly obeyed your laws? There may be a few individuals who have followed your commands, but not an entire people.

Then the angel Uriel, who had been sent to me, spoke and said, "Are you so troubled by this world that you wish to understand the ways of the Most High?"

I replied, "Yes, my Lord."

The angel continued, "I have three challenges for you. If you can answer one of them, I will show you the truth about why there is evil in the world."

I said, "Please tell me."

He said, "Weigh fire on a scale, measure the wind, or bring back a day that has already passed."

I answered, "No one born of man could do these things. Why do you ask me something impossible?"

Then he said, "What if I had asked you how many chambers are in the ocean, how many springs flow from underground, or how many paths lead to heaven? Could you answer? You would say you have never gone to the depths of the sea, nor to the underworld, nor have you risen to heaven. But I did not ask about those things. I asked about fire, wind, and time—things you live with every day. And yet, you cannot explain them."

Then he said, "If you cannot understand what happens around you, how can you expect to understand the ways of the Most High? His ways are beyond human understanding. A mortal person living in a corrupt world cannot grasp the wisdom of an eternal God."

When I heard this, I fell to the ground and said, "It would have been better if we had never been born. Now we live in sin, we suffer, and we do not even understand why."

The angel answered, "Let me tell you a story.

"Once, the trees of the forest gathered together and said, 'Let's go to war against the sea so that it will move back and give us more land to grow.'

"But the waves of the sea also gathered together and said, 'Let's go to war against the trees so that we can take their land for the waters.'

"But in the end, fire came and burned the trees, and the sand rose up to stop the waves. Both plans were useless.

"If you had been the judge, who would you say was right and who was wrong?"

I replied, "Neither plan made sense. The trees were meant to stay on the land, and the sea was meant to hold its waves."

Then the angel said, "You have judged correctly! So why do you not judge yourself in the same way? Just as trees belong to the land and waves belong to the sea, people on earth can only understand what happens on earth. Only those in heaven can understand the things above."

I answered, "Then tell me, my Lord, why was I given the ability to think if I cannot understand these things? I am not asking about what is above the heavens. I only want to understand what happens around me every day.

"Why has Israel been handed over to other nations? Why have your chosen people fallen into the hands of those who do not believe in you? Why has the holy Law been ignored? Why have the covenants written for us disappeared?

"We pass from this world like insects, and our lives are as brief as a breath. We do not deserve your mercy, but what will you do for the sake of your great name, which we carry? That is what I want to understand."

The Answer

The angel answered me, "If you live long enough, you will see and be amazed because this world is quickly passing away. It cannot last forever, especially since it is full of suffering and pain. It is not strong enough to hold the great rewards promised to the righteous.

"The evil you asked me about has already been planted, but its time to be harvested has not yet come. Until the evil is removed, the good cannot fully grow. From the very beginning, one seed of evil was planted in Adam's heart, and look at how much sin it has produced over time. And it will continue until the time for the harvest arrives.

"Now, think for yourself—how much wickedness has grown from that one seed? So imagine when countless seeds of goodness are planted—how great will the harvest be?"

I asked, "How long will it take for this to happen? Our lives are short and full of trouble."

The angel replied, "You cannot rush ahead of the Most High. You want things to happen quickly for yourself, but God is waiting for the sake of many. Even the souls of the righteous, who are waiting in their resting places, ask, 'How much longer must we wait? When will we receive our reward?'

"The angel Remiel answered them, saying, 'You must wait until the full number of righteous people has been reached.'

"The Holy One has measured the world carefully. He has counted the times and set the seasons in place. He will not change anything or rush the process until everything is complete."

I said, "Lord, but the world is full of sin! Are the righteous being delayed because of all the evil that people do?"

The angel answered, "Go ask a pregnant woman if she can hold her baby inside her after nine months."

I replied, "No, my Lord, she cannot."

Then he said, "The underworld and the resting places of souls are like a mother's womb. Just as a woman cannot hold back her child when it is time to give birth, the world cannot delay what has been set in motion. When the time comes, everything that has been waiting since the beginning will be revealed."

I asked, "If I have found favor in your eyes, please show me something else. Has more time already passed than what remains? I know about the past, but I do not know what is still ahead."

The angel said, "Stand to my right, and I will show you."

I stood and saw a huge fire burning fiercely. When the flames died down, I saw that smoke remained. Then I saw a heavy storm cloud filled with water, pouring down a violent rain. But after the storm passed, a few drops of water were still left behind.

Then the angel said, "Think about what you saw. Just as the fire was much greater than the smoke, and the rain was heavier than the drops that remained, so too has most of time already passed. Only a little remains."

The Signs Which Precede the End (IV. 51-V.13)

I asked him, "Will I live to see those days? Who will be alive when these things happen?"

He answered, "I can tell you some of the signs you asked about, but I was not sent to tell you about your life. I do not know the answer to that."

"As for the signs," he continued, "the time is coming when people on earth will be overcome with fear. Truth will be hidden, and faith will disappear. Sin and shamelessness will be worse than anything you see now or have heard about in the past.

"The land you see now will become unstable and abandoned. If the Most High allows you to live, you will see the land in total confusion after three days.

"Strange things will happen. The sun will shine at night, and the moon will be bright during the day. Trees will drip blood, and stones will speak. People will be in chaos, and the air will change.

"A ruler will rise whom no one expects, and birds will leave their usual places.

"The sea of Sodom will be full of fish, and a mysterious voice will be heard at night, surprising everyone.

"Cracks will open in many places, and fire will keep bursting out. Wild animals will leave their habitats, and strange things will happen to pregnant women—some babies will not develop fully before birth.

"Fresh water will become salty. Friends will suddenly turn against each other in battle.

"Wisdom will disappear, and understanding will be hidden. People will search for them but will not find them.

"Sin and shamelessness will spread everywhere. People will ask their neighbors, 'Have you seen anyone doing what is right?' But the answer will always be 'No.'

"During that time, people will hope but never receive what they wish for. The land will struggle to produce food, and hard work will not lead to success.

"I was told to share these signs with you. But if you pray again as you have been doing, and fast for seven more days, you will hear even greater things."

The Conclusion of The Vision (V. 14-19)

I woke up, shaking all over, and felt so weak that it seemed like my life was slipping away.

But the angel who had been speaking to me reached out, helped me stand, and gave me strength.

On the second night, Phaltiel, the leader of the people, came to me and asked, "Where have you been? Why do you look so troubled? Don't you know that you have been chosen to care for Israel while they are in captivity?

"Get up and eat something so you don't abandon us, like a shepherd leaving his flock to be attacked by wolves!"

But I told him, "Leave me alone. Do not come near me for seven days. After that, you can return, and I will explain everything to you."

After I said this, he left.

Vision II

The Prayer of Ezra 1

I fasted for seven days, crying and mourning, just as the angel Ramiel had told me to do. After the seven days, my heart was still troubled, and I felt overwhelmed with thoughts. But then, my soul was filled with understanding, and I began to pray again, speaking to the Most High with deep pleading.

I said, "Lord, from all the trees in the world, you have chosen one vine. From all the lands, you have picked one special place. From the deep waters of the sea, you have set aside one river. Among all the flowers, you have chosen one. From all the cities ever built, you made Zion your holy place.

"Among all the birds, you have named one dove as special. From all the animals, you have chosen one sheep. And from all the nations, you have brought one people close to you, giving them the law you approved of and loved.

"But now, Lord, why have you given the chosen one over to so many? Why have you allowed your special people to be scattered among the nations? Those who reject your commandments have trampled on the ones who believed in your covenant. If you were so angry with your people, shouldn't you have punished them yourself instead of letting others destroy them?"

After I finished speaking, the angel who had appeared to me before returned and said, "Listen to me, Ezra, and I will explain. Look at me, and I will give you understanding."

I replied, "Speak, my lord."

The angel asked, "Are you so troubled about Israel? Do you love them more than the One who created them?"

I said, "No, my lord! But my heart is in pain, and I suffer every moment because I want to understand God's judgment."

The angel said, "You cannot understand."

I asked, "Why not, my lord? Why was I even born if I cannot understand these things? Why wasn't my mother's womb my grave so I would not have to see the suffering of Jacob and the pain of Israel's descendants?"

The angel answered, "If you can count the people who have not yet been born, gather the raindrops that have fallen, or make dead flowers bloom again—if you can open doors that have never been unlocked, control the winds that are held back, or show me the face of someone you have never seen—then I will tell you what you want to know."

I replied, "Lord, only the One who does not live among men can know these things. I am weak and foolish—how could I possibly understand what you ask?"

The angel said, "Just as you cannot do any of these things, you also cannot understand God's judgment or the depth of the love He has promised His people."

Then I asked, "Lord, you made promises to those who will live at the end of time. But what about the people who came before us, those who live now, and those who will come after us?"

The angel answered, "God's judgment is fair to all. There is no advantage for those who came first, and no disadvantage for those who come last."

I asked, "But couldn't you have created everyone all at once so that we could all be judged together?"

The angel answered me and said, "Creation does not move faster than its Creator. If everything were created all at once, the world would not be able to support it."

I replied, "But you just told me that all creation will be brought back to life at the same time. If that is true, then why couldn't everything have been made at once from the start?"

The angel responded, "Ask a woman who is pregnant, 'If you carry ten children, why don't you give birth to them all at once?' Demand that she deliver them all at the same time."

I answered, "She cannot, my lord. She can only give birth at the right time."

He said, "In the same way, I have made the earth like a womb, where people come into the world at different times. Just as a newborn child cannot give birth and an old woman can no longer bear children, I have set the world in order."

Then I asked, "Since you have explained this to me, let me ask another question. The city of Zion, which you spoke about—does she still have her strength, or is she growing old?"

The angel said, "Ask a mother who has given birth, and she will tell you. Say to her, 'Why are the children you bear now smaller than those who came before?' She will answer, 'Children born when I was young were strong, but those born when I was old are weaker because my body has aged.'

"In the same way, if you look at the people of today, you will see that they are smaller and weaker than those who lived before them. And the people who come after you will be even weaker because creation itself is growing old, and its strength is fading."

Then I said, "Please, my lord, if I have found favor in your eyes, tell me who will bring the end of the world."

The angel answered, "The beginning was in the hands of men, but the end will come by my own hands.

"Before the land of the world was formed, before the winds blew, before thunder was heard, before lightning flashed, before the garden of Paradise was planted, before the flowers bloomed, before the mighty forces of nature moved, before the angels gathered, before the sky stretched high, before the foundations of Zion were set, before the present time was measured, before people began to sin, and before those with faith were chosen—I had already planned it all. Everything that exists was made by my own hand, not by anyone else."

Then I asked, "What does it mean that time will be divided? When will the first age end and the next one begin?"

The angel answered, "From Abraham to Abraham. Abraham had Isaac, and Isaac had Jacob and Esau. When they were born, Jacob held Esau's heel.

"Esau represents the end of one time, and Jacob represents the beginning of another. Just as a man's life starts with his hand and ends with his heel, time moves in the same way. There is nothing more you need to understand, Ezra."

The Signs of The Last Time and The End

I replied, "Lord, if I have found favor in your eyes, please show me the full meaning of the signs you have revealed to me in part during the night."

The angel answered, "Stand up on your feet, and you will hear a loud voice. If the ground beneath you shakes while you listen, do not be afraid. The voice will be speaking about the end of time, and even the foundations of the earth will understand this message. They will shake and tremble because they will know that their time is changing."

As I listened, I stood up and heard a voice like the sound of rushing waters. It said:

"The time is coming when I will visit those who live on the earth and hold the wicked accountable for their actions. When the suffering of Zion is finished and this world is about to come to an end, I will show these signs:

- The books of judgment will be opened in the sky, and everyone will see my decision.
- One-year-old children will be able to speak.
- Pregnant women will give birth after only three or four months, and their babies will live and move.
- Fields that were once empty will suddenly grow crops.
- Storehouses full of food will suddenly be empty.
- A loud trumpet will sound, and everyone will hear it and be afraid.
- Friends will turn against each other like enemies, and the earth will be shocked by what happens.
- Springs of water will stop flowing for three hours.

Anyone who survives these events will witness my salvation and the end of the world. They will see those who have been taken up and have never experienced death. The hearts and minds of the people will be changed, and they will think differently.

- Evil will be wiped away, and lies will disappear.
- Faith will grow, and corruption will be defeated.
- Truth will finally appear after being hidden for so many years."

As the voice spoke, I felt the ground beneath me begin to shake little by little.

The Conclusion of The Vision (VI. 30-34)

The angel said to me, "I have come to reveal these things to you tonight. If you pray and fast for another seven days, I will show you even greater things.

Your voice has been heard by the Most High. The Mighty One has seen your sincerity and the holiness you have shown since your youth. That is why He has sent me to reveal these things to you.

Take courage and do not be afraid! But do not be quick to judge the past, or you may be judged in the final days."

Vision III

(VI. . 35-IX. 25)

Introduction (VI. 35-37)

After this, I cried and went without food for seven days so I could complete the three weeks I had been told to do. On the eighth night, I felt troubled again, and I spoke to the Most High because my heart was heavy with emotion, and my soul felt like it was burning inside me.

I said, "Lord, you spoke at the beginning of creation and commanded the heavens and earth to exist, and your Word completed the work. Your Spirit moved over everything, while darkness covered all, and no human voice had yet been heard. Then you commanded a ray of light to come from your treasures so that all your works could be seen.

On the second day, you created the sky to separate the waters, placing some above and some below. On the third day, you gathered the waters into one area, revealing dry land. You set aside part of the land to be used for farming and ordered it to produce food. Immediately, all kinds of fruit grew—too many to count—each with its own taste. Beautiful flowers bloomed in many shapes, and trees appeared, each different from the other. Their scents were beyond description. All of this happened on the third day.

On the fourth day, you commanded the sun, moon, and stars to shine and serve mankind, whom you were about to create. On the fifth day, you told the waters to bring forth all kinds of creatures—birds, fish, and animals—so that they would show your wonders. Though the waters had no life of their own, they produced living things. You also created two great creatures: one you called Behemoth, and the other Leviathan. You separated them because the sea could not hold them both together. Behemoth was placed in a dry area with a thousand mountains, while Leviathan was kept in the waters. You have saved them for when you decide they will be used as food.

On the sixth day, you commanded the earth to bring forth land animals, crawling creatures, and livestock. Over all of these, you placed Adam as their ruler, and from him, we, your chosen people, have come.

I have told you all of this, Lord, because you said you created the world for our sake. But as for the other people who also come from Adam, you said they are nothing, like a drop of water from a bucket or spit on the ground. And yet now, these same people rule over us and oppress us! But we, your chosen people, your firstborn and beloved ones, have been given into their hands.

If this world was created for us, then why do we not rule over it? How much longer must we endure this?"

After I finished speaking, the angel who had visited me before came again and said, "Stand up, Ezra, and listen to the words I have come to tell you."

I said, "Speak, my Lord."

He answered, "Imagine a vast sea that is wide and endless, but there is only a narrow entrance, like a river. If someone wants to enter the sea, see it, and take control of it, they must first go through the narrow path. If they do not pass through the narrow way, how can they ever reach the open sea?"

Listen to this example: There is a city built in a wide valley, filled with many good things. However, the entrance to the city is narrow and placed on high ground. On one side, there is fire, and on the other, deep waters.

A single narrow path runs between them, just wide enough for one person to walk at a time. If someone is meant to inherit that city, how can they receive it unless they first pass through the dangers that block the way?

I answered, "That makes sense, my Lord!" And he said to me, "This is also true for Israel. I created the world for them, but when Adam disobeyed my commands, everything was cursed. That is why life in this world is now full of suffering, pain, and danger. But the next world will be wide, peaceful, and full of everlasting goodness.

If people don't go through struggles and hardships, they won't be able to receive what has been prepared for them.

So why are you upset that people are weak and mortal? Why do you focus only on what is happening now instead of looking at what is to come?

I answered, "Lord, you said in your Law that good people will inherit these blessings, but the wicked will be destroyed. The righteous endure hardships because they hope for a greater reward, but the wicked suffer too—and they never get to see the reward!"

He replied, "You are not wiser than God or greater than the Most High! Many will perish because they rejected my Law, which I gave them so they could live. I told them what to do to avoid punishment, but they refused to listen. Instead, they chose to follow empty and worthless thoughts, turned away from the truth, and even denied that the Most High exists.

They rejected his Law, broke his promises, ignored his commands, and refused to acknowledge his works.

So, Ezra, empty things belong to those who are empty, but those who are full will receive what is full."

The Temporary Messianic Kingdom and The End of The World (VP. 26-[44])

The time is coming when the signs I told you about will happen. The bride, who has been hidden, will appear like a great city, and what was once cut off will be seen again.

Anyone who survives these troubles will witness my wonders.

My son, the Messiah, will be revealed, along with those who are with him, and they will bring joy to those who remain for thirty years. After that time, my son, the Messiah, will die, along with all people who are still living. Then, for seven days, the world will return to the silence it had in the beginning, and no one will be left.

After those seven days, the world will wake up again, and corruption will be destroyed. The earth will give up the dead, the dust will release those buried in it, and the hidden places will return the souls that were placed there. The Most High will sit on the throne of judgment, and the end will come. Mercy will be no more, and patience will be gone. Only my judgment will remain. Truth will stand, and faith will grow.

The good and evil deeds of all people will be revealed. The spirit of torment will appear, but so will the place of rest. The fires of

Gehenna will be shown, but across from it will be the paradise of peace.

Then the Most High will say to the nations, "Look at what you have rejected! See the one you refused to serve and the commands you ignored! Look before you—on one side, rest and joy; on the other, fire and suffering!" That is how he will speak on the Day of Judgment.

On that day, the sun, moon, and stars will not shine. There will be no clouds, lightning, or thunder. The wind, water, and air will not move. There will be no darkness, no morning or evening, no summer or winter. There will be no heat, cold, frost, hail, dew, or rain. Time itself—day and night—will no longer exist. There will be no lamps, no torches, no light, except for the shining glory of the Most High. From that light, people will see what has been decided for them.

There will be a waiting period, like a week of years, but this is part of the plan. I have revealed it only to you.

I answered and said, "Lord, I have said before and I say again: Blessed are those who have lived and obeyed your commands. But what about those I prayed for? Is there anyone who has lived without sin? Is there anyone born who has never disobeyed your command?

Now I see that few will find joy in the world to come, but many will face suffering.

For inside us is an evil heart that leads us away from the right path. It drags us into corruption and shows us the way to death. It leads us far from life, and this has happened not just to a few, but to nearly everyone who has ever lived."

The angel answered me and said, "Listen, Ezra, and I will explain things to you again.

The Most High did not create just one world, but two. You asked why there are so few righteous people, so let me give you an example.

If you had a few precious stones, would you compare them to piles of lead and clay?

I asked, "How does that make sense, Lord?"

And he replied, "Ask the earth, and she will tell you. Speak to her, and she will explain. The earth produces gold, silver, copper, iron, lead, and clay. But silver is more common than gold, copper is more common than silver, iron is more common than copper, lead is more common than iron, and clay is the most common of all. Now, tell me, what is more valuable—the rare things or the abundant ones?"

I answered, "Lord, the things that are rare are the most precious, while the things that are common have little value."

Then he said to me, "Now think about what you just said. People rejoice more over a small amount of something valuable than over an abundance of something worthless. The same is true for my judgment. I take joy in the few who live righteously because they bring glory to my name. But I do not grieve over the many who perish, because they are like a breath of air, like smoke that vanishes, like a flame that burns out and disappears."

Then I cried out, "Oh, earth, what have you done? You have brought forth people who are now doomed to destruction! If human intelligence comes from the dust like everything else, then it would have been better if the dust had never existed, so that intelligence would never have been born.

But now, intelligence grows within us, and that is why we suffer—because we understand what is happening, yet we still perish!

Let the human race mourn, but let the animals of the field rejoice! Let all who are born weep, but let the cattle and sheep celebrate! It is better for them than for us, because they do not fear judgment, they do not know punishment, and they were never promised life after death.

So what good is it for us to live, only to suffer?

Every person is born into sin, burdened with wrongdoing from the start. If we did not have to face judgment after death, it would be better for us!"

Then the angel answered me and said, "When the Most High created the world, Adam, and all who came after him, he also prepared judgment and everything that comes with it.

Now think about your own words—you said that intelligence grows within people. That is exactly why they must face judgment! They had understanding, yet they chose to do evil. They received commandments but did not follow them. The Law was given to them, but they rejected it.

What will they say on the day of judgment? How will they defend themselves when the time comes?

For so long, the Most High has been patient with the people of this world—not for their sake, but because the right time had to come."

The State of The Soul Between Death And Judgement VII.

I asked, "Lord, if I have found favor with you, please tell me this: After we die and our souls leave our bodies, do we rest until the time comes when you renew the world, or do we immediately face torment?"

He answered, "I will explain this to you, but do not associate yourself with those who rebel or suffer punishment. You have stored up good deeds with the Most High, and your reward will be revealed at the appointed time.

As for death, here is what happens: When a person's time to die comes, as determined by the Most High, their spirit leaves their body and returns to the One who gave it. First, the soul acknowledges the glory of God.

But if the soul belonged to someone who denied God, rejected His ways, or hated those who feared Him, it does not enter a place of rest. Instead, it immediately suffers torment in seven ways:

1. They realize they have disobeyed the law of the Most High.
2. They understand they can no longer repent or do good deeds to save themselves.
3. They see the reward given to those who were faithful.
4. They recognize the punishment waiting for them in the end and regret not following God when they had the chance.
5. They look upon the peaceful rest of the righteous souls, knowing they will never experience it.
6. They see the suffering prepared for them and know it is unavoidable.
7. Worst of all, they are consumed with shame and fear as they see the glory of the Most High, the one they disobeyed in life, and they know they will face His judgment.

But for those who have followed the ways of the Most High, this is what happens when their time comes:

While they were alive, they faithfully served God, enduring struggles and hardships to keep His commandments. Because of this, when they die, they experience seven great joys:

1. They rejoice in the victory of overcoming evil desires and staying on the path of life.

2. They see the suffering of the wicked and realize they have been saved from it.
3. They hear the Most High Himself testify that they have been faithful to His law.
4. They rest peacefully in the chambers of the righteous, guarded by angels, and they see the glorious future that awaits them.
5. They celebrate their escape from the struggles of the world, knowing they now inherit an eternal reward.
6. They see how their faces will shine like the sun and how they will become like the stars, never again to experience corruption.
7. Above all, they feel fearless and confident as they prepare to stand before the One they served in life, knowing they will be honored and rewarded by Him.

These are the paths of the righteous after death, but the disobedient will only know suffering. Their souls do not enter a place of peace but remain in torment, grieving and regretting their choices in seven ways.

Then I asked, "After the soul leaves the body, is there a time when it sees these things you have described?"

He answered me, saying, "For seven days after death, souls are free to see these things I have told you about. After that, they are gathered into their resting places."

Then I asked, "If I have found favor with you, please tell me—on the Day of Judgment, will the righteous be able to pray for the wicked or ask the Most High to have mercy on them? Can fathers pray for their sons, or sons for their fathers? Can brothers, relatives, or friends plead for one another?"

He replied, "Since you have been shown favor, I will explain this as well. The Day of Judgment is final and will reveal the truth to all. Just as no one can take another's sickness, hunger, or suffering in their place—whether a father, a son, a master, or a friend—so too, on that day, no one can pray for another. Each person will carry the weight of their own righteousness or sin."

I said, "But Lord, in the past, Abraham prayed for the people of Sodom. Moses prayed for our ancestors in the wilderness when they sinned. Joshua prayed for Israel after Achan's wrongdoing. Samuel prayed in the time of Saul, and David for the suffering of the people. Solomon prayed for those in the Temple. Elijah prayed for rain and even for the dead to come back to life. Hezekiah prayed for the people when Sennacherib attacked. Many have prayed for others before—so why can't this happen on the Day of Judgment?"

He answered, "This world has an end, and God's glory does not remain in it forever. That is why the strong have prayed for the weak. But the Day of Judgment is different. It is the end of this world and the beginning of the next, which will never die. In that world, corruption will be gone, wickedness will be destroyed, and faithlessness will be no more. Righteousness will flourish, and truth will shine. On that day, no one will be able to help those who are condemned, just as no one can harm those who are saved."

The Promises of Future Felicity Only Mock A Sin-Stained Race (VII. [N6]-[131])

I answered, "This is my final thought: It would have been better if the earth had never created Adam, or if, when he was created, you had taught him not to sin.

"What good has it done for people to live in suffering, only to die and face punishment? Oh, Adam, what have you done? You were the one who sinned, but the consequences were not just yours—they became ours too, because we came from you!

"What is the point of being promised eternal life when our actions lead to death? What good is it to be given hope that never fades, when we are trapped in misery? What benefit is there in knowing that places of safety and healing exist if we have lived wickedly?

"We are told that the glory of the Most High will protect those who lived righteously, but we chose to follow evil instead. Paradise, filled with fruit that never withers, joy, and healing, has been revealed, but we cannot enter it because of our sinful ways. The faces of the holy will shine brighter than the stars, but our faces will be covered in darkness.

"When we were alive and doing wrong, we never stopped to think about the suffering we would face after death."

He answered me, saying, "This is the challenge that every person faces in life. If they fail, they will suffer, just as you have said. But if they overcome it, they will receive the rewards I have described.

"This is the same message that Moses gave to the people while he was alive. He told them, 'I have set before you today life and death, good and evil. Choose life, so that you and your children may live.'

"But they refused to listen to him. They ignored the prophets who came after him. And even now, they do not believe what I am telling them.

"So, there should be no sadness over their destruction, just as there is joy over those who have chosen life."

Will The Merciful and Compassionate One Suffer So Many To Perish? (VII. [132.]-VIII. 3)

I answered and said, "Lord, I understand that the Most High is called compassionate because he shows kindness even to those who have not yet been born.

"He is called gracious because he welcomes those who turn to his law.

"He is patient because he gives us, his creation, time even when we sin.

"He is generous because he prefers to give rather than take.

"He is full of mercy because he offers kindness not just to those who are alive now, but also to those who have passed and those who are yet to come.

"If he did not show such great mercy, neither the world nor its people would be able to survive.

"He is generous because, in his goodness, he lessens the punishment of sinners—otherwise, hardly one in ten thousand people would still be alive.

"He is also a judge, because if he did not forgive those he created and overlook many of their sins, only a very small number of people would remain."

VIII. R. And He Answered and Said to Me: This World Hath the Most High Made for The Sake of Many, But That Which Is to Come for The Sake of Few.

I will share a parable with you, Ezra. If you ask the earth what it produces more of—ordinary clay or precious gold—it will tell you that common dust is far more abundant than gold. The same is true for this world: many people are created, but only a few truly live.

I answered and said, "Let my soul seek understanding, and let my heart gain wisdom! We come into this world without choosing to, and we leave it without wanting to. We are only given a short time to live.

Lord, if you allow me, I will pray to you. Please give us new hearts that can grow good and lasting things so that those who are weak and temporary can have life.

You are the one and only Creator, and we are the work of your hands, as you have said. In the womb, you form our bodies and shape our features. You protect us in warmth and water for nine months until we are born.

The mother's body nourishes the child as you intended, providing milk to help it grow. You guide us with your kindness, sustain us with justice, and teach us through your wisdom. You give us life, and when the time comes, you take it away.

But if you destroy what you took such care to create, what was the purpose of making us in the first place?

I am speaking about all people, but even more, I am speaking about your chosen people. It breaks my heart to see their suffering, and I grieve for the people of Israel.

Now, I will begin to pray for myself and for others, because I see how much we have sinned in this world. I have also heard about the coming judgment. So, please, Lord my God, listen to my prayer as I speak to you.

The Seer's Prayer for The Divine Compassion on His People,

The prayer of Ezra before he was taken up:

Lord, you live forever, your throne has no limits, and your glory is beyond understanding. The mighty ones stand in fear before you, and at your command, they turn into fire and wind. Your words are true and unchanging, your commands are powerful, and your voice is terrifying.

You can dry up the deep waters with just a look, and your rebuke can make mountains melt. Your truth is clear for all to see. Please listen to your servant's voice, hear my prayer, and pay attention to my words.

As long as I live, I will speak, and as long as I have understanding, I will call out to you. Please do not focus on the sins of your people, but remember those who have served you faithfully. Do not hold onto the foolish actions of the wicked, but think of those who have kept your promises, even when they were treated with shame.

Do not judge those who have done evil before you, but remember those who have feared you with sincerity. Do not destroy those who have acted like animals, but instead look upon those who have followed the light of your law. Do not stay angry with those who have behaved worse than beasts, but love those who have always trusted in your glory.

We, along with those who came before us, have done wrong and acted foolishly. Yet because of our sins, you are called the Compassionate One. If you are willing to show mercy to us, even though we have no good deeds to offer, you will be known as the Gracious One.

The righteous who have done good already have their rewards stored with you. But what is a person that you should be so angry with them? Why be so harsh with a mortal race?

The truth is, there is no one who has not sinned, no one who has lived without doing wrong. So, Lord, let your kindness be known by showing mercy to those who have no good works to offer.

The Divine Reply

And he answered me, saying: Some of the things you have said are correct, and they will happen as you have spoken.

I do not focus on those who do evil, their death, their judgment, or their destruction. Instead, I take joy in the creation of the righteous, in their lives, and in the rewards they will receive.

For as you have said, so it will be.

Mankind Is Like Seed Sown (VIII. 41-45)

Just as a farmer plants many seeds and grows many plants, not all of them survive or take root. In the same way, not everyone who is born will live.

And I answered, "Lord, if I have found favor in your sight, may I ask something? A farmer's seeds will not grow unless they receive rain at the right time, and too much rain can even destroy them. But people are different—they were made by your own hands, in your own image, and you created everything for their sake. How can they be compared to seeds?

Please, Lord, have mercy on your people. Show kindness to your creation, for they belong to you, and you are full of compassion."

The Final Reply: Let the Seer Contemplate the Lot o The Blessed Which He Is Destined to Share (VIII. 46 62)

And he answered me, saying:

"The things of this world belong to those who live in it now, and the things of the future are for those who will come later. You cannot love my creation more than I do. But you have compared yourself to the wicked too many times—do not do this!

However, you will be honored before the Most High because you have humbled yourself, as you should, and have not tried to place yourself among the righteous. Because of this, you will be given even greater honor.

In the end, those who live on the earth will suffer greatly because of their pride. But instead of worrying about them, think about yourself and ask about the rewards of those who are like you.

For you, Paradise is open, the Tree of Life is planted, and the future world is prepared. A place of joy is waiting, a city has been built, and a peaceful rest has been set aside. All that is good has been made complete, and wisdom has reached its fullness. Evil has been locked away, sickness will no longer exist, death will disappear, the grave will be forgotten, and pain will be no more.

In the end, the treasures of life will be revealed. So do not ask again about those who will perish, because they were given freedom, but they rejected the Most High.

They scorned his law and worked to remove his ways from the

earth. They even mistreated his faithful ones and said in their hearts, 'There is no God,' even though they knew they would die one day.

Just as the rewards I told you about are waiting for the righteous, suffering and torment are also prepared for the wicked. The Most High never wanted people to be destroyed, but they dishonored his name and refused to be grateful. They rejected the one who gave them life.

And so, my judgment is near. This is something I have not revealed to many people, but only to you and a few like you."

IX. The Signs of The End Reviewed (VIII. 63-IX. 63.

And I said, "Lord, you have shown me many signs of what will happen in the last days, but you haven't told me when it will happen."

He answered, "Pay close attention and think carefully. When you see that some of these signs have already happened, you will know that the Most High is preparing to visit the world He created.

When you notice earthquakes, crowds in chaos, people plotting against each other, leaders struggling for power, and rulers in confusion, understand that these are the events the Most High spoke about long ago.

Just like everything in the world has a clear beginning and end, the times set by the Most High are also known. Their beginnings come with warnings and signs of power, and their end will bring judgment and more signs.

Anyone who survives or escapes—whether through their actions or through their faith—will be safe from the dangers I have described. They will see my salvation in the land I have set apart forever.

But those who ignored my ways will be shocked, and those who rejected and abandoned them will suffer.

All who refused to recognize me while they were alive, even when I was kind to them, and all who rejected my laws while they had the freedom to follow them, will realize the truth after death.

The Fewness of The Saved Further Justified

So don't focus on how the wicked will be punished. Instead, think about how the righteous will live—the ones for whom this world was created.

I replied, "I keep saying, and I will say it again, that more people will be lost than saved. It's like comparing the vast ocean waves to a tiny drop of water."

He answered, "Just as the land determines the kind of seeds that grow, and flowers decide their own colors, and work produces different smells, everything follows a pattern. The farmer's field matches his efforts.

Before people even existed, I had already prepared a place for them to live. No one could stop me—because no one was there yet.

Now that they are here, living in a stable world with everything they need and a law they cannot fully understand, they have become corrupt in their actions. I looked at my world and saw that it was ruined. I saw my creation and realized it was in danger because of the way people behaved.

So I decided to spare only a few. I saved one grape from a whole bunch and one plant from an entire forest.

Let the many be destroyed, since they were created for nothing. But let my chosen ones remain, for they are precious and took great effort to bring into existence.

Conclusion Of The Vision

If you separate yourself for seven more days—but this time, do not fast—go to an open field full of flowers, where no buildings have been built. Eat only the flowers from the field. Do not eat meat or drink wine, only the flowers.

Pray with deep devotion to the Most High, and I will come to you and speak with you.

Introduction (IX. 26-28)

I went to the field called Arpad, just as he told me, and sat among the flowers. I ate the plants from the field, and they satisfied me.

After seven days, as I lay on the grass, I felt something stir in my heart again, just like before. Then my mouth opened, and I began to speak to the Most High.

The Glory of The Law and Israel: A Contrast (IX. 29-37)

And I said, "Lord, you truly showed yourself to our ancestors in the wilderness when they left Egypt. They traveled through a barren land where nothing grew and no one had ever lived.

You spoke to them, saying, 'Listen to me, Israel, and pay attention, descendants of Jacob! I am planting my Law within you, and it will grow and produce good things. Through it, you will be honored forever.'

But our ancestors received the Law and did not follow it. They were given commandments but did not obey them. The Law itself never faded because it belongs to you, but those who received it perished because they failed to live by it.

It's like how the earth receives seeds, or the sea holds a ship, or a container is filled with food—these things may be used up or destroyed, but what holds them remains. But for us, it has been different.

We received the Law, yet because of our sins, we are destroyed along with our hearts that once accepted it.

Still, your Law does not fade away. It remains in its full glory."

The Vision of The Disconsolate Woman (IX. 38--X. 24)

As I was thinking about these things, I looked up and saw a woman on my right. She was crying loudly, deeply distressed, and sighing in sorrow. Her clothes were torn, and she had thrown dust on her head in grief.

I stopped thinking about my own concerns and turned to her, asking, "Why are you weeping? What is causing you such deep pain?"

She replied, "Please, my lord, allow me to cry freely and continue to grieve, for my heart is filled with sorrow, and I feel completely broken."

I said, "Tell me what has happened to you."

She answered, "I was unable to have children. For thirty years, I was married but could not conceive. Every single day and night, I prayed to the Most High, begging for a child.

Then, after those thirty years, God finally heard my prayers. He saw my suffering, understood my pain, and blessed me with a son.

I was overjoyed. My husband, my neighbors, and I all celebrated and praised the Mighty One. I raised my son with great effort and love.

When he grew up, I arranged for him to be married and planned a joyful wedding feast. But on the night of his wedding, as he entered his new home, he suddenly collapsed and died.

In my grief, I put out the lights, and the people in my town came to comfort me. But I remained silent, waiting until the next day and through the night.

Once everyone was asleep and thought I was resting too, I got up in the darkness, ran away, and came to this field where you see me now.

I have decided that I will never return to the city. I will not eat or drink, but I will continue to mourn and fast until I die."

I let go of my own thoughts and, in my frustration, said to her, "You are being more foolish than any other woman! Can't you see the suffering around us? Do you not realize what has happened to all of us?

Look at Zion, the mother of us all—she is in deep sorrow and has been humiliated beyond measure.

Yes, you grieve for your one son, but we grieve for an entire world in mourning.

Ask the earth, and she will tell you—she has witnessed the birth of countless people, and every one of them, from the beginning of time until now, has passed away. Many more will come, and they too will face destruction.

Who, then, should grieve more? You, for your one son, or the earth, which has lost countless lives?

And if you say, 'My grief is different because I lost the child I carried in my womb, the one I gave birth to and raised with love and hardship,'

Remember that the earth, just like a mother, has brought forth all of humanity. Every person who has ever lived was born from the earth, and all return to it.

So now, keep your sorrow within you. Face your pain with strength, and endure the hardship that has come upon you!"

If you accept the Most High's judgment as fair, then one day, you will be reunited with your son and be honored among women.

So go back to the city, return to your husband.

But she replied, "I will not go back. I will not return to the city or to my husband. I will stay here and die."

I continued speaking to her, saying, "No, don't do this! Instead, think about Zion's suffering and take comfort in Jerusalem's sorrow.

Look at what has happened—our sacred places are destroyed, our altars torn down, our Temple ruined. Our worship has ended, our songs have stopped, and our joy has faded. The light of our lamp has gone out, the ark of the covenant has been taken, and our holy places have been defiled.

The name we carry has been dishonored, our leaders have been shamed, our priests burned, and our Levites taken as prisoners. Our young women have been violated, our wives have been abused, our prophets captured, our watchmen scattered, our children enslaved, and our strongest warriors have been brought low.

And worst of all—the very symbol of Zion's glory has been taken away and handed over to those who hate us.

So let go of your overwhelming grief. Turn to the Mighty One so He may have mercy on you, and the Most High may give you rest from all your suffering."

Sion's Glory; The Vision of The Heavenly Jerusalem (X. 25 28)

As I was speaking with her, her face suddenly began to shine brightly, like a flash of lightning. I was terrified and too afraid to go near her, my heart filled with shock and confusion.

As I tried to understand what was happening, she suddenly let out a loud, terrifying cry, so powerful that the entire earth seemed to shake at the sound of her voice.

Then, as I looked again, she was gone. In her place, I saw a great city with strong, deep foundations. Fear took hold of me, and I cried out,

"Where is the angel Uriel, who has been with me since the beginning? He is the one who led me into this overwhelming experience. Now, I feel lost, my body weak, and my prayers seem worthless."

The Vision Interpreted (X. 29-57)

As I lay on the ground, feeling as if I were dead, the angel who had spoken to me before returned. He saw me there, weak and confused, and took my right hand. He helped me stand up and gave me strength. Then he asked,

"What is troubling you? Why are you so disturbed? Why is your mind so overwhelmed?"

I replied, "Because you left me! I did everything you told me—I went into the field, and I saw something beyond my understanding. I cannot explain it."

The angel said, "Stand up, and I will help you understand."

I said, "Please, my lord, speak to me, but do not leave me again, or I fear I will die too soon.

I have seen things I do not understand, and I have heard things that confuse me. Is my mind deceiving me? Am I only seeing a dream?

I beg you, my lord, explain this terrifying vision to me."

The angel answered, "Listen carefully, and I will teach you. I will reveal what you are afraid of, because the Most High has shared many secrets with you.

He has seen your righteous heart, how deeply you grieve for your people, and how much you mourn for Zion.

Now, here is the meaning of what you saw.

The woman who appeared to you in mourning—the one you tried to comfort—

She was not just a woman. She was Zion, the city you now see being built before you.

When she spoke of her son's misfortune, this is what it means:

The woman you saw is Zion itself, which has now become a great city.

When she told you she had been barren for thirty years, it represents the three thousand years before offerings were ever made there.

Then, after three thousand years, Solomon built the city and offered sacrifices in it. That is when the 'barren woman' finally bore a son.

When she spoke of raising her son with hardship, that represents the building and growth of Jerusalem.

When she said her son entered his wedding chamber and died, this symbolizes the fall and destruction of Jerusalem.

You saw how she mourned for her children, and you tried to comfort her.

Now, the Most High has seen how deeply you grieve for Zion, how your heart is truly broken for her suffering.

Because of this, He has allowed you to see her future glory and her true beauty.

That is why I told you to wait for me in the field, where no houses were built.

I knew the Most High was about to reveal these things to you.

That is also why I brought you to a place with no buildings—because no human structure could remain where the City of the Most High was about to be revealed.

But do not be afraid. Let your heart be at peace.

Go forward and see the light of Zion's glory. Look at the greatness of her buildings, as far as your eyes can see.

Then, you will hear as much as your ears can take in.

You are blessed above many others, for the Most High has chosen you among only a few."

Transition To the Fifth Vision {X. 58-59)

But tomorrow night, you must stay here.

The Most High will show you a vision of the events that will take place on earth in the last days.

Vision V

(X. 60-XII. The Vision (X. 6o---XII. Ja)

I stayed there that night, just as I was told.

The next night, I had a vision. I saw a huge eagle rising from the sea. It had twelve wings and three heads. As I watched, it spread its wings over the whole earth, and the winds from the sky blew around it. Clouds gathered toward it.

Then, I saw that smaller wings grew out of its large wings, but they were weak and unimportant. The eagle's heads remained still, but the middle head was larger than the others, even though it too was resting.

Then, the eagle commanded its wings to rule over the earth and its people. I saw that everything beneath the sky became subject to it, and no creature on earth resisted.

The eagle then stood on its claws and spoke to its wings, saying, "Go and rule over the earth. But rest now—do not all rise at once. Wake up at different times, and leave the heads for last."

I noticed that the eagle's voice did not come from its heads but from the middle of its body. I counted its small wings—there were eight.

Then, I saw one wing rise from the right side and rule over the earth. But after a while, it disappeared completely, leaving no trace.

Then, a second wing rose and ruled for a long time, but eventually, it too was destroyed like the first.

I heard a voice speaking to the second wing:

"You who have ruled for so long, listen! No ruler after you will hold power for as long as you did—not even for half as long."

Then, a third wing rose and ruled, just like the others, but it was also destroyed. The same happened to each of the wings—one after another, they ruled and then fell.

Later, I saw some of the smaller wings on the right side try to rule the earth. Some succeeded but were quickly destroyed. Others rose but never gained power.

Eventually, all twelve wings were gone, along with two of the small wings. All that remained of the eagle was its three resting heads and six small wings.

Then, I saw two of the small wings separate from the others and move under the right head, while the remaining four stayed in place.

I watched as these four small wings tried to rise to power. One of them succeeded but was quickly destroyed. The second tried, but it fell even faster than the first.

The last two small wings also thought they would rule, but before they could, something happened.

The middle head, which had been resting, suddenly woke up. It was larger than the other two heads.

It joined with the other two heads and turned on the last two small wings, devouring them before they could take power.

Then, the middle head ruled over the entire earth. It treated the people harshly and had more power than all the wings before it.

But suddenly, the middle head was destroyed, just like the wings had been.

Now, only the two remaining heads ruled over the earth and its people. Then, I saw the head on the right devour the head on the left, leaving only one.

Then, I heard a voice say, "Look ahead, Ezra, and see what happens at the end."

I looked and saw a lion coming out of the forest. It roared loudly, and then I heard it speak with a human voice. It spoke to the eagle, saying:

"Listen, eagle! I will tell you what the Most High says.

Are you not the one who remains from the…"

The Most High said, "I created four great beasts to rule over the world, and through them, the end of time would come.

You, the fourth beast, have risen above all the others before you. You have ruled over the earth with cruelty and have brought suffering to the whole world. You have lived among people for a long time, deceiving them and ruling unfairly.

You have stolen from the poor and mistreated those who are honest. You have harmed the humble, hated those who do what is right, and loved those who are deceitful. You have destroyed the homes of the successful and torn down the walls of those who never wronged you.

Because of your arrogance, your sins have reached the Most High.

Now, He has looked upon the times, and they have come to an end. The days He set in place have been fulfilled.

So, you, the eagle, will be completely destroyed—your great wings, your small and wicked wings, your cruel heads, your sharp claws, and your entire hateful body.

The earth will finally be at peace, free from violence. It will no longer suffer but will look forward to the judgment and mercy of its Creator."

Then, after the lion spoke these words to the eagle, I saw the last remaining head suddenly destroyed.

Then, the two wings that had tried to take power rose up to rule, but their rule was short and filled with chaos.

I watched as they too were destroyed, and the entire body of the eagle was burned. The earth stood in shock at what had happened.

The Interpretation Of The Vision (XII. 3B-39)

I woke up in great distress and fear and said to myself, "This is happening to me because I have been seeking to understand the ways of the Most High.

Now, I feel completely drained, my spirit is weak, and I have no strength left because of the overwhelming fear I experienced last night.

But I will pray to the Most High, and He will give me the strength to endure."

Then I prayed, "Lord, if I have found favor in your eyes, if you have truly blessed me above many others, and if my prayers have reached you,

Then strengthen me and help me understand this vision I have seen. Please explain it to me so that my soul may find peace.

Did you not choose me to reveal the end of times and the completion of these events?"

Then He answered me, saying, "This is the meaning of the vision you saw:

The eagle that rose from the sea represents the fourth kingdom, the same one your brother Daniel saw in his vision. However, it was not explained to him as I am explaining it to you now.

A time will come when a kingdom will rise on the earth, more powerful and terrifying than all the kingdoms before it.

Twelve kings will rule over it, one after another. But the second king will rule longer than the others.

This is the meaning of the twelve wings you saw.

As for the voice that spoke, not from the eagle's head but from the middle of its body, this means that during the middle of that kingdom's reign, it will face many divisions and nearly collapse. However, it will not fall at that time but will recover and continue to rule.

The eight small wings that grew under its larger wings represent eight kings who will rise within the kingdom. Their reigns will be short and their rule unstable. Two of them will die early, four will be saved for the final period, and two will remain until the very end.

The three resting heads you saw represent three powerful kings that the Most High will bring at the end of this kingdom's time. They will change many things and will oppress the world and its people with even more cruelty than those before them.

That is why they are called the heads of the eagle—because they will bring its final wickedness before its end.

The one large head that was destroyed means that one of these kings will die naturally, though he will still suffer.

As for the two remaining kings, they will be killed by the sword. One will destroy the other, but in the end, he too will fall by the sword.

The two wings that moved to the head on the right side represent those whom the Most High has set apart for the final moments. Their rule will be short and filled with chaos, just as you saw.

The lion that came from the forest, roaring and speaking to the eagle, correcting it for its wickedness—this represents the Messiah.

The Most High has kept Him for the final days. He will come from the line of David and will confront the rulers of this world. He will call them out for their evil, expose their corruption, and show them their sins.

He will bring them before God for judgment while they are still alive. After He rebukes them, He will destroy them.

But He will show mercy to my people, those who have remained faithful and stayed within my borders. He will bring them joy until the final Day of Judgment arrives, just as I have told you before.

This is the vision you saw, and this is its meaning.

You alone have been chosen to understand this mystery of the Most High.

So write down everything you have seen in a book and keep it in a safe place.

Teach it only to those among your people who are wise and have the understanding to keep these secrets.

But you must remain here for seven more days, for the Most High will reveal even more to you."

Conclusion Of the Vision (XII. 39B-48)

Then he left me.

When the people saw that seven days had passed and I had not returned to the city, they all gathered together—young and old—and came to me. They asked,

"What have we done wrong? How have we sinned against you that you have abandoned us and chosen to stay out here?

You are the last prophet left to us, like the last bunch of grapes from the harvest, like a light in the darkness, like a safe harbor for a ship caught in a storm.

Are all the troubles we have suffered not enough? Must we also lose you?

If you leave us too, then it would have been better if we had died when Zion was burned.

We are no better than those who perished there."

Hearing this, I cried out loudly and wept. Then I answered them,

"Have courage, Israel! Do not be discouraged, House of Jacob.

The Most High remembers you, and the Mighty One will not forget you forever.

I have not abandoned you, and I never will.

I came here to pray for the destruction of Zion and to ask for mercy for our Sanctuary, which has been humiliated."

Vision VI

(THE MAN FROM THE SEA)

(XIII. I-58)

The Vision (XIII. 1-13a)

After seven days, I had a vision in the night. I saw a powerful wind rise from the sea, stirring up massive waves.

Then, from deep within the sea, a figure like a man appeared. He flew through the sky on the clouds of heaven. Everywhere he looked, the earth trembled before him. When he spoke, those who heard his voice melted away like wax in a fire.

Then, I saw an enormous army gathering from all directions, coming together to fight against the man who had risen from the sea.

I watched as he carved out a massive mountain for himself and stood on top of it. I wanted to see where the mountain had come from, but I couldn't find its source.

Even though the armies were terrified, they still prepared to fight him.

But he didn't raise his hand, pick up a spear, or use any weapon of war. Instead, he opened his mouth, and streams of fire, powerful winds, and burning coals poured out.

The fire, wind, and stormy flames combined and struck the army with overwhelming force. The entire multitude was instantly burned up, leaving nothing behind but ashes and smoke. I was shocked at what I saw.

Then, the man came down from the mountain and called out to another group of people, who peacefully came to him.

Many people approached him—some happy, some sad. Some were in chains, while others brought those who were to be offered.

At this point, I woke up in great distress. I prayed to the Most High and said,

"Lord, from the beginning, you have shown me these great wonders. Even though I am unworthy, you have answered my prayers.

Now, please reveal the meaning of this vision to me.

I fear for those who will be left alive in those days, but even more for those who will not survive.

Those who do not live to see these events will mourn what they have missed.

But those who do survive will suffer greatly, facing extreme dangers and hardships, as these visions have shown.

Yet, it is better to endure and witness these things than to disappear like a cloud and never see how the end of time unfolds."

Then He answered me, saying,

"I will explain your vision and also tell you about the people you have asked about.

As for those who survive and those who do not—here is the meaning:

Whoever endures the dangers of that time will be saved, as will those who have remained faithful and lived righteously before the Most High.

So know this—those who survive will receive more blessings than those who have died."

L. THE APOCALYPSE OF EZRA THE INTERPRETATION OF THE VISION (XIII. 25-53A)

Here is what your vision means:

The man you saw rising from the sea is the one the Most High has been keeping for a long time. He will come to save creation and bring to safety those who remain.

The fire, storm, and breath that came from his mouth—without the use of weapons—destroying the army that gathered to fight him, means this:

The time is coming when the Most High will rescue those on earth. A great terror will fall upon the people.

Nations will turn against each other—cities against cities, places against places, people against people, and kingdoms against kingdoms.

When these signs take place, just as I told you before, my Son will be revealed—the same man you saw rising from the sea.

When people hear his voice, they will abandon their battles and conflicts.

Then, as you saw, an uncountable number of people will gather to fight against him.

But he will stand on Mount Zion, and Zion itself will appear before everyone—fully prepared and built, just as you saw the mountain that was not made by human hands.

My Son will confront these people for their wickedness, like a storm striking against them.

He will reveal all their evil deeds and the punishment they are about to face.

Then, like a blazing fire, he will destroy them effortlessly—by the Law of the One who is like fire.

The peaceful people he gathered afterward are the nine and a half tribes who were taken from their land during the time of King Josiah.

The Assyrian King Salmanassar captured them and took them to the other side of the Euphrates River, exiling them to a distant land.

But they made a decision among themselves—to leave behind the other nations and go to a place where no people had ever lived before.

There, they hoped to keep the Law that they had failed to follow in their own land.

They traveled through the narrow paths of the Euphrates, and the Most High performed miracles for them.

He held back the river's waters until they had all crossed safely.

Their journey was long—it took a year and a half—until they reached a place called Arsaph, at the farthest edge of the world.

They have lived there until the end times.

But when their return is near, the Most High will once again stop the flow of the Euphrates, so they can cross back.

That is why you saw a great gathering of people coming in peace.

The people from your own nation who remain within my sacred land will also survive.

When my Son destroys the gathered armies, he will protect those who are left, and then he will show them many wonders.

I then asked, "Lord, why did I see the man coming from the sea?"

He answered, "Just as no one can search the deep sea or fully know what lies beneath, no one on earth can see my Son or those with him—except in the time of his coming.

This is the meaning of your vision."

Transition To the Seventh Vision (XIII. 53B-58)

These things have been revealed to you alone

because you have set aside your own concerns and dedicated yourself to understanding the ways of the Most High. You have sought to learn the truths of the Law.

You have lived wisely and have treated understanding as your guide, like a mother who teaches her child.

That is why I have shown you these things—because the Most High has a reward for you.

In three days, I will speak to you again and reveal even greater wonders.

Then I went out into the field, walking for a long time, praising the Most High for the amazing things He has done throughout history.

I thanked Him for how He controls time and everything that happens within it.

And I stayed there for three days.

Vision VII

(Ch. Xiv)

Ezra's Commission (XIV. 1-17)

After this, as I was sitting under an oak tree, a voice suddenly came from a bush in front of me.

It called out, "Ezra, Ezra!"

I answered, "Here I am!" and stood up.

The voice said, "I revealed myself from a bush and spoke with Moses when my people were enslaved in Egypt.

I sent him to lead my people out of Egypt, through the wilderness, and up to Mount Sinai. I kept him near me for many days,

showing him great wonders, revealing the secrets of time, and explaining how everything would come to an end.

I told him that some of these words must be kept secret, while others should be shared.

Now I say the same to you, Ezra.

The signs I have shown you, the visions you have seen, and the explanations you have heard—keep them in your heart and hide them.

Because you will be taken away from this world, and you will stay with my Son and others like you until the end of time.

This world is growing old, and its time is almost over.

So, put your life in order, warn your people, comfort those who are struggling, and guide the wise.

Let go of this temporary life,

free yourself from the concerns of this world, stop worrying about death, and cast away all weakness.

Do not let these thoughts trouble you—leave these times behind!

Because the troubles you have already seen will be followed by even greater ones.

As the world ages, evil will increase, and suffering will spread among its people.

Truth will fade, and lies will take over.

Look! The great eagle you saw in your vision is already on its way."

Ezra Prays for Inspiration (XIV. R8-26)

I answered and said, "Lord, please let me speak!

I will do as you have commanded and warn the people who are alive now. But what about those who have not yet been born? Who will warn them?

The world is covered in darkness, and its people have no light.

The Law has been burned, and no one knows the works you have done or what you are about to do.

If I have found favor in your eyes, Lord, send your Holy Spirit to me. I will write down everything that has happened in the world from the very beginning—everything written in your Law—so that people may find the right path and those who live in the last days will know the way.

He answered and said, "Go, gather your people, and tell them not to look for you for forty days.

Prepare many writing tablets, and take with you Seraia, Daria, Shelemia, Helkana, and Shiel—these five men, because they are skilled in writing quickly.

Come here, and I will place a light of understanding in your heart that will not go out until you have finished writing.

When you are done, you will share some of what you write with everyone, but some you must keep hidden and give only to the wise.

Tomorrow at this time, you will begin writing."

Ezra's Last Words (XIV. 27-36)

I went and did as I was commanded, gathering all the people together and saying to them:

"Listen, Israel, to these words.

Our ancestors were once strangers in the land of Egypt, but they were rescued from there.

They received the Law of life, but they did not follow it. And just like them, you have also broken it.

You were given the land of Zion as an inheritance, but you and your ancestors sinned and did not follow the ways that Moses, the servant of the Lord, commanded you.

So the Most High, who is a fair judge, took away what had been given to you for a time.

Now, you are suffering here, and your brothers and sisters have been scattered to a faraway land.

But if you turn back to the truth, discipline your hearts, and live rightly, you will be saved. And after death, you will receive mercy.

For after death, there will be judgment, and we will live again. Then, the names of the righteous will be honored, and the sins of the wicked will be exposed.

Until then, do not come near me or try to find me for forty days."

The Restoration of The Scriptures (XIV. 37-48)

I took the five men, just as I was commanded, and we went into the field, where we stayed.

The next day, a voice called out to me, saying, "Ezra, open your mouth and drink what I give you."

I opened my mouth and saw a full cup coming toward me. It looked like it was filled with water, but its appearance was like fire. I took the cup and drank from it.

As soon as I drank, my heart was filled with understanding, my mind overflowed with wisdom, and my spirit held onto knowledge. My mouth opened, and I could not stop speaking.

The Most High gave the five men the ability to understand, and they wrote down everything I spoke, using writing they had never known before.

I stayed there for forty days. During the day, I spoke, and they wrote. At night, they ate bread, but I remained awake, never stopping.

By the end of forty days, we had written ninety-four books.

When the forty days were over, the Most High spoke to me and said,

"The first twenty-four books you have written, make public. Let both those who are worthy and those who are not read them.

But keep the other seventy books hidden, and give them only to the wise among your people.

For in these books are the deep secrets of understanding, the sources of wisdom, and the path to knowledge."

I followed the command exactly, in the seventh year, during the sixth week, five thousand years after creation, on the twelfth day of the third month.

Conclusion Of the Book (XIV. 49-50)

After writing everything down, Ezra was taken away to be with others like him.

He was forever known as the Scribe of the Knowledge of the Most High.

This was the conclusion of Ezra's first account.

Third Baruch

Chapter One

I, Baruch, was deeply troubled and filled with sorrow as I thought about the suffering of my people. My heart ached, and I mourned over how King Nebuchadnezzar had been allowed by the Almighty to destroy the holy city. I struggled to understand why this had happened to us. Crying out, I said, "Lord, why did you allow your vineyard to be ruined? Why have you let it be destroyed? Why did you punish us in this way instead of disciplining us differently? Instead, you handed us over to these nations that now mock us, saying, 'Where is their God?'"

As these thoughts overwhelmed me, tears streamed down my face, and my heart grew heavier with sorrow. But as I wept, pouring out my grief, a vision appeared before me. I saw an angel of the Lord descending with a bright and powerful presence. He came near and spoke to me, saying, "Man of God, greatly loved, do not let your heart be so burdened over the fate of Jerusalem. Listen to the words of the Almighty, for He has sent me to reveal His divine plan to you."

The angel continued, "This is what the Lord God Almighty says: Your prayer has been heard, Baruch. The Most High has received your cries and your sorrow. He has seen the pain in your heart and has listened to your words."

Hearing this, I began to feel a sense of calm, though my mind was still filled with questions. The angel, sensing my uncertainty, said, "Do not trouble yourself by trying to understand everything right now. The ways of God are beyond what you can see. I have been sent to show you things far greater than what you have asked. These are truths beyond your understanding at this moment."

With deep respect, I replied, "As surely as the Lord lives, I will not question or complain if you reveal these mysteries to me. And if I fail to keep this promise, may God judge me on the day of reckoning."

Then the angel, speaking with both authority and kindness, said, "Come, Baruch. Follow me, and I will show you the hidden and sacred things of the Most High. Open your heart to receive these truths, for they will bring light to your soul and wisdom to your spirit."

At that moment, I resolved to listen, ready to receive the knowledge that the Lord was about to reveal through His messenger. My sorrow began to change into anticipation as I followed the angel, preparing to understand the deep and eternal wisdom of the Almighty.

Chapter Two

The angel took me with him to where the heavens were firmly set in place. There, I saw an enormous river—so wide that no one could cross it, not even the foreign nations that God had created. Then, he led me further and took me up to the first heaven, where we arrived at a massive door. He turned to me and said, "Let's go inside."

As we entered, it felt as if we were flying, traveling a distance that would have taken thirty days to walk.

Inside, I saw a vast plain stretching across the heaven. There were beings living there, but they looked different from humans. Their faces resembled cattle, they had horns like deer, feet like goats, and their lower bodies were covered in wool like sheep.

I, Baruch, turned to the angel and asked, "Please tell me, how thick is this heaven we passed through? How wide is it? And what is this plain you have shown me? I want to share this with the people on earth."

The angel, whose name was Phamael, answered, "The door you saw is the gateway to heaven. Its thickness is as great as the distance from earth to heaven, and the width of the plain you saw is just as vast."

Then the angel said, "Come with me, and I will show you even greater mysteries."

I asked him, "Please tell me, who are these people?"

He replied, "They are the ones who built the great tower, trying to fight against God. Because of their actions, the Lord removed them from the earth."

Chapter Three

The angel of the Lord took me up to the second heaven and showed me a door that looked just like the first one. He said, "Let's go inside." So we entered, traveling as if we were flying, covering a distance that would take sixty days to walk.

Inside, I saw another vast plain, and it was filled with people. But their appearance was strange—they had faces like dogs and feet like deer.

I asked the angel, "Who are these people?"

He answered, "These are the ones who planned to build the great tower. They forced many men and women to make bricks for the construction. Among them was a woman who was forced to keep working even while she was giving birth. They did not allow her to stop, so she gave birth while making bricks. She wrapped her baby in her cloak and continued working.

When the Lord appeared to them, He confused their languages. By that time, they had already built the tower up to a height of 463 cubits. Then they took a tool and tried to drill through the sky, saying, 'Let's see if heaven is made of clay, copper, or iron.'

When God saw what they were doing, He did not let them continue. Instead, He struck them with blindness and caused their speech to become confused. That is why they are as you see them now."

Chapter Four

I, Baruch, said, "Lord, you have already shown me incredible and amazing things. Now, I ask you, for the Lord's sake, to show me everything else."

The angel replied, "Come, let's go further." We traveled together, covering a distance that would take about 185 days to walk.

He led me to a vast plain where I saw a massive serpent that looked like it was made of stone. Then, he showed me Hades, a place that was dark, empty, and unclean.

I asked, "What is this dragon, and what is the creature surrounding it?"

The angel explained, "This dragon devours the bodies of those who lived wicked lives. It feeds on them. Hades works the same way—it consumes and never stops. Each day, it takes about a cubit of water from the sea, yet the sea never runs dry."

I asked, "How is that possible?"

The angel said, "Listen carefully. The Lord created 360 rivers, and the three largest ones are the Alphias, the Aburos, and the Gerikos. These rivers constantly flow into the sea, keeping it from ever running out."

Then I said, "Please show me the tree that caused Adam to go astray."

The angel answered, "That tree is actually a vine, planted by the angel Samail. It made the Lord angry, and He cursed both Samail and the vine. That's why God commanded Adam not to touch it. The devil, out of jealousy, used it to deceive Adam."

I asked, "If the vine was so dangerous and brought a curse upon Adam, why does it still serve an important purpose today?"

The angel replied, "That is a good question. When God sent the flood to destroy all living things, including 409,000 giants, the water rose 15 cubits above the highest mountains. The flood reached Paradise and wiped out every plant and flower. However, a small branch of the vine was carried by the waters and left on the earth.

"When the flood ended and the land appeared again, Noah came out of the ark and began planting everything he found. Among them, he discovered the vine branch. He wasn't sure what it was and prayed for guidance. That's when I came to him and explained its origin.

"Noah asked, 'Should I plant this, or should I destroy it? Since Adam was cursed because of it, will I also bring God's anger upon myself if I plant it?' He was unsure, so he prayed for forty days, pleading with God to tell him what to do.

"Finally, God sent the angel Sarasel with a message. The angel told Noah, 'Plant the vine. The Lord says this: What was once bitter will become sweet, what was cursed will become a blessing, and its fruit will represent the blood of God. Just as humanity was condemned through it, through Jesus Christ, Emmanuel, it will now offer a way back into Paradise.'

"Remember this, Baruch: Just as Adam was cursed and lost God's glory because of this vine, people today also fall further from God when they drink too much wine. By overindulging, they bring judgment upon themselves and prepare for eternal punishment.

"Nothing good comes from drinking in excess. Those who drink too much commit terrible sins: brothers turn against brothers, fathers lose compassion for their sons, children stop respecting their parents, and because of drunkenness, all kinds of evil arise—murder, adultery, immorality, lying, stealing, and many other sins. Truly, no good comes from it."

I, Baruch, said, "Lord, you have already shown me amazing and powerful things. Now, I ask you, for the sake of the Lord, to show me everything else."

The angel replied, "Come, let's go further." We traveled together, covering a distance that would take about 185 days to walk.

He led me to a vast plain where I saw a huge serpent that looked as if it were made of stone. Then, he showed me Hades—a dark, empty, and unclean place.

I asked, "What is this dragon, and what is the creature surrounding it?"

The angel explained, "This dragon feeds on the bodies of those who lived sinful lives. It grows stronger by consuming them. Hades works the same way—it takes from the world, but never runs out. Every day, it absorbs a cubit of water from the sea, yet the sea never shrinks."

I asked, "How is that possible?"

The angel said, "Listen carefully. The Lord created 360 rivers, and the three largest ones are the Alphias, the Aburos, and the Gerikos. These rivers flow constantly into the sea, keeping it from ever running dry."

Then I said, "Please show me the tree that led Adam astray."

The angel answered, "That tree is actually a vine, planted by the angel Samail. It angered the Lord, and He cursed both Samail and the vine. That is why God commanded Adam not to touch it. The devil, out of jealousy, used it to deceive Adam."

I asked, "If this vine caused such harm and was cursed by God, why does it still have such an important purpose?"

The angel replied, "That is a good question. When God sent the great flood to destroy all living things, including 409,000 giants, the water rose 15 cubits above the highest mountains. The flood even reached Paradise, wiping out every plant and flower. However, a small branch of the vine was carried by the waters and left on the earth.

"When the flood ended and the land reappeared, Noah left the ark and began planting the things he found. Among them, he discovered the vine branch. He was unsure what it was and prayed for guidance. That's when I came to him and explained its origin.

"Noah asked, 'Should I plant this, or should I destroy it? Since Adam was cursed because of it, will I also bring God's anger upon myself if I plant it?' He was uncertain, so he prayed for forty days, asking God to tell him what to do.

"Finally, God sent the angel Sarasel with a message. The angel told Noah, 'Plant the vine. The Lord says this: What was once bitter will become sweet, what was cursed will become a blessing, and its fruit will represent the blood of God. Just as humanity was condemned through it, through Jesus Christ, Emmanuel, it will now offer a way back into Paradise.'

"Remember this, Baruch: Just as Adam was cursed and lost God's glory because of this vine, people today also fall further from God when they drink too much wine. By overindulging, they bring judgment upon themselves and prepare for eternal punishment.

"Nothing good comes from drinking excessively. Those who do become reckless and commit terrible sins: brothers turn against brothers, fathers lose compassion for their children, children stop respecting their parents, and because of drunkenness, all kinds of evil arise—murder, adultery, immorality, lying, stealing, and many other wrongs. Truly, nothing good comes from it."

Chapter Five

I, Baruch, turned to the angel and asked, "Lord, may I ask you something?"

The angel replied, "Ask whatever you wish."

I continued, "You told me that the serpent drinks a cubit of water from the sea each day. Can you tell me how big its stomach is?"

The angel answered, "Its stomach is as large as Hades itself. It is so massive that it stretches as far as a group of 300 men could throw a heavy stone."

Then he said, "Come with me, and I will show you things even greater than this."

Chapter six

The angel took me to the place where the sun rises.

He showed me a chariot pulled by four horses, with fire blazing beneath it. A man sat on the chariot, wearing a crown made of flames. Around the chariot stood forty angels, and in front of it ran a massive bird—so large that it looked as big as nine mountains.

I asked the angel, "What is this bird?"

He answered, "This is the guardian of the world."

I asked again, "How does this bird protect the world? Please explain it to me."

The angel said, "This bird travels with the sun as it moves. It spreads its wings to absorb the sun's fiery rays. If it didn't do this, the heat would be too strong, and no living creature—human or animal—would survive. That is why God created this bird for this purpose."

Then the bird spread its wings, and I saw large letters written on its right wing. The letters covered a huge space, about the size of a threshing floor, around 4,000 measures wide. They shined like gold.

The angel said, "Read what is written."

So I read the letters, and they said: "Neither the earth nor the heavens can hold me, but the wings of fire carry me."

I asked the angel, "What is the name of this bird?"

He replied, "It is called the Phoenix."

I asked, "What does it eat?"

He answered, "It feeds on manna from heaven and the dew of the earth."

I asked again, "Does it produce waste?"

The angel replied, "Yes, it excretes a worm, and from this, cinnamon is formed. This cinnamon is used by kings and rulers. But wait, and you will see the glory of God."

While the angel was still speaking, a sudden thunderclap shook the ground beneath us.

I asked, "What is this sound?"

The angel explained, "The angels are opening the 365 gates of heaven, allowing light to separate from darkness."

Then I heard a voice say, "Light-giver, bring splendor to the world!"

At that moment, I also heard the bird's call and asked, "What is that sound?"

The angel answered, "This is the cry that wakes the roosters on earth. Just as people signal each other, the rooster announces the start of the day. As the angels prepare the sun, the rooster crows to let the earth know morning has come."

Chapter Seven

I asked, "Where does the sun begin its journey after the rooster crows?"

The angel replied, "Listen, Baruch, everything I have shown you so far is within the first and second heavens. But in the third heaven, the sun moves through and spreads its light across the world. Be patient, and you will see the glory of God."

As the angel spoke, I saw the bird appear again. At first, it looked small, but it grew larger and larger until it returned to its full size.

Following the bird, I saw the sun shining brightly, surrounded by angels who carried it. A magnificent crown rested on the sun, and its light was so intense that we couldn't look directly at it.

At the same moment, the phoenix spread its wings wide. The brilliance of the scene was overwhelming, and I was filled with such fear that I turned away and hid under the angel's wings.

The angel reassured me, saying, "Do not be afraid, Baruch. Stay here, and soon you will see the sun as it sets."

Chapter Eight

He led me toward the west, and as the sun neared the horizon, I once again saw the great bird flying ahead, guiding the sun. The sun followed closely behind, surrounded by angels. When it reached its resting place, I watched as the angels removed the crown from its head. The bird, exhausted from its journey, let its wings droop as if weighed down by fatigue.

Curious about what I had seen, I asked, "Why do the angels take the crown off the sun's head? And why does the bird look so tired?"

The angel explained, "At the end of each day, four angels carry the sun's crown to heaven so that it can be renewed. This is necessary because, as the sun travels over the earth, its rays and crown become tainted. Each day, they must be cleansed and restored."

I then asked, "How do the sun's rays become unclean?"

The angel replied, "As the sun shines on the earth, it witnesses the sins of humanity. These include acts of immorality, stealing, violence, idolatry, drunkenness, murder, jealousy, gossip, deceit, and many other things that offend God. Because of this, the sun's light becomes stained and must be purified daily."

Wanting to understand more, I asked, "Why is the bird so worn out?"

The angel answered, "The bird is exhausted because it spends the entire day protecting the earth from the full heat of the sun. It spreads its wings to absorb the intense rays so that they don't burn everything below. Without its constant effort, as I told you before, no living thing would be able to survive the sun's scorching heat."

Chapter Nine

When night arrived, the moon and stars appeared in the sky. I, Baruch, turned to the angel and asked, "Lord, please explain this to me as well. Where does the moon go when it disappears, and what path does it follow?"

The angel replied, "Be patient, and soon you will see and understand."

The next day, I saw the moon. It looked like a woman seated in a chariot with wheels. In front of the chariot were oxen and lambs, and many angels traveled alongside it.

I asked, "Lord, what are the oxen and lambs?"

The angel answered, "These are also angels."

I then asked, "Why does the moon sometimes appear larger and at other times smaller?"

The angel explained, "Listen, Baruch. The moon was created by God to be beautiful and unique. However, during Adam's first sin, the moon gave its light to Samael when he took the form of the serpent. Instead of hiding its brightness, it shined even more. This angered God, so He reduced its light and shortened its days."

I asked, "Why doesn't the moon shine all the time, but only at night?"

The angel replied, "Just as servants do not speak freely in the presence of a king, the moon and stars cannot shine in the presence of the sun. The stars remain in their places, but their light is overpowered by the sun's brightness. Meanwhile, the moon is safe, but its light fades because of the sun's intense heat and brilliance."

Chapter Ten

After the archangel had taught me all these things, he led me to the third heaven. There, I saw a vast, endless plain, and in the center was a peaceful lake of clear water. Surrounding the lake were many birds of all kinds, but they were unlike any birds I had ever seen on earth. Among them, I noticed a crane as large as an ox. All of the birds were magnificent, far greater than anything found in the world below.

I turned to the angel and asked, "What is this plain? What is this lake? And why are so many birds gathered around it?"

The angel replied, "Listen carefully, Baruch. This plain, which holds many hidden mysteries, is where the souls of the righteous come together. Here, they live in peace and form choirs to praise the Lord."

He continued, "The water in this lake is what the clouds draw up to bring rain to the earth. That rain helps plants grow and produce fruit."

I then asked, "And what about these birds?"

The angel answered, "These birds never stop singing praises to the Lord."

I, Baruch, then said, "Lord, why do people say that rain comes from the sea?"

The angel explained, "Some rain does come from the sea and from water sources on the earth. But the rain that helps crops grow comes from this place. From now on, understand that what people call the 'dew of heaven' also comes from here."

Chapter Eleven

The angel led me from that place to the fifth heaven. When we arrived, I saw that the gate was closed. I asked, "Lord, will the gate be opened so we can enter?"

The angel replied, "We cannot go in until Michael, the one who holds the keys to heaven, arrives. Be patient, and you will see the glory of God."

Suddenly, a loud noise like thunder filled the air. I asked, "Lord, what is that sound?"

The angel answered, "It is the archangel Michael descending to collect the prayers of people on earth."

Then a voice called out, "Let the gate be opened!" At that moment, the gate swung open with a sound as powerful as a thunderclap.

Michael appeared, and the angel with me stepped forward, bowing low and saying, "Greetings, commander of all the heavenly armies."

Michael replied, "Greetings to you as well, our brother and messenger of God's revelations to the righteous."

After they exchanged greetings, they stood together. Then, I saw Michael holding a massive bowl. It was so large that its depth stretched from heaven to earth, and its width reached from north to south.

Amazed, I asked, "Lord, what is that bowl that the archangel Michael is carrying?"

The angel answered, "This bowl holds the virtues and good deeds of the righteous. Michael gathers them and presents them before God in heaven."

Chapter Twelve

As I was speaking with them, I saw angels approaching, each carrying baskets filled with beautiful flowers. They brought these baskets to Michael and handed them over. Curious, I turned to the angel beside me and asked, "Lord, who are these angels, and what are they carrying?"

He answered, "These are the angels responsible for overseeing different regions and nations."

I watched as Michael took the baskets from the angels and emptied the flowers into the large bowl he was holding. Then, the angel explained, "The flowers being poured into the bowl represent the good deeds and virtues of the righteous."

As I looked closer, I noticed other angels arriving, but their baskets were not completely full. These angels seemed troubled and hesitant, standing back instead of stepping forward. It was clear they hadn't gathered enough to fill their baskets.

Michael saw their hesitation and called out, "Come forward, you angels, and bring whatever you have collected."

They obeyed, but as they poured what little they had into the bowl, both Michael and the angel beside me looked deeply saddened. The contributions from these angels were not enough to fill the bowl completely.

Chapter Thirteen

Then, more angels arrived, crying and trembling with fear. They said, "Look at us, Lord! We have become stained and darkened because we were assigned to serve evil people. Please, we beg you, remove us from their presence."

Michael replied, "You cannot leave them, for we must not let the enemy claim victory. But tell me, what do you wish?"

The angels answered, "Michael, our commander, we ask you to take us away from them. We can no longer bear being among these wicked and corrupt people. They have no goodness in them—only greed and wrongdoing.

"We have never seen them enter a place of worship, seek guidance from spiritual leaders, or do anything righteous. Instead, wherever there is murder, they are involved. Wherever there is immorality, theft, lies, jealousy, drunkenness, violence, or idolatry, they are at the center of it all. Their actions are evil, and they continue in their wickedness without regret. Please, release us from them."

Michael turned to them and said, "Wait here while I seek the Lord's will to find out what should be done."

Chapter Fourteen

At that moment, Michael left, and the large doors closed tightly behind him. A deep, powerful sound, like rolling thunder, echoed through the heavens.

I turned to the angel beside me and asked, "What is that loud noise?"

The angel replied, "That is Michael presenting the good deeds of people before God."

Chapter Fifteen

At that moment, Michael returned, and the gate opened. He carried oil with him.

For the angels who had brought full baskets, he filled them with oil and said, "Take this and give a hundredfold reward to our friends—those who have worked hard and done good. Those who have planted well will harvest well."

To the angels who had brought half-filled baskets, he said, "Come and receive your reward based on what you brought, and deliver it to the people on earth."

Then, speaking to both groups—the ones with full baskets and those with half-filled ones—he said, "Go and bless our friends. Tell them that the Lord says: 'You have been faithful with little, so I will give you even more. Enter into the joy of the Lord.'"

Chapter Sixteen

Then he turned to those who had brought nothing and said, "Do not be sad, and do not weep, but do not abandon the people on earth either.

Since they have angered me with their actions, go and make them jealous. Stir their frustration and turn their hearts against those who are not even a nation, against people who lack understanding.

Send swarms of caterpillars and locusts, rust and grasshoppers. Strike them with hail, lightning, and fury. Bring punishment through the sword and death, and send demons to trouble their children.

For they refused to listen to my voice. They ignored my commands and did not follow them. Instead, they rejected my instructions, turned away from my places of worship, and insulted the priests who spoke my words to them."

Chapter Seventeen

As he finished speaking, the door closed, and we stepped away.

The angel then led me back to the place where my journey had first begun.

When I came to my senses, I praised God for allowing me to witness such incredible things.

And to you, my brothers who read these revelations, give glory to God, so that He may also honor us—now and forever, for all eternity! Amen.

Sibylline Oracles (Apocalyptic Portions)

Introduction

The Sibylline Oracles are a collection of ancient prophetic writings created over several centuries. They combine ideas from Greek, Jewish, and early Christian traditions. These oracles are linked to the Sibyls, legendary prophetesses of the ancient world. Written in Greek poetry, they were used to spread religious and political messages. Their long history reflects the mix of different cultures and beliefs in the Mediterranean from ancient times to the early Middle Ages.

The idea of the Sibyl originally came from ancient Greece, where people believed that one prophetess revealed divine messages. Over time, different Sibyls appeared across the Mediterranean, including the famous Cumaean Sibyl, who, according to Roman mythology, advised the hero Aeneas. The Sibylline Oracles are not the same as the original Roman Sibylline Books, which were official prophetic texts lost long ago. Instead, the surviving oracles are a mix of writings from different sources, created between the 2nd century BCE and the 7th century CE. They combine Greek and Roman myths with Jewish, Gnostic, and early Christian ideas, forming a unique collection of prophecy.

The oracles contain 12 main books, totaling over 4,000 lines. The Sibyl speaks in the first person and mostly describes the future. The content varies, including visions of the end of the world, moral lessons, and prophecies about final judgment. Some sections use acrostics, where the first letters of each line spell out hidden messages, showing the detailed writing style of the authors.

These oracles were written during times of political and religious struggles. Jewish writers used the Sibyl's voice to challenge oppressive rulers and promote belief in one God, blending Jewish ideas into the prophecies. As Christianity spread, Christian writers added references to Jesus and Christian teachings, reshaping the oracles to support their religious messages. This mix of traditions made the oracles useful for connecting with pagan audiences, giving Christian beliefs more credibility by presenting them through a respected prophetic figure.

The oracles were passed down through handwritten copies from the 14th to the 16th centuries. These manuscripts show signs of editing over time, and scholars have grouped them based on content and history. Early Christian leaders, such as Theophilus of Antioch, Clement of Alexandria, and Lactantius, often quoted the oracles, showing their influence in early Christian writings.

The Sibylline Oracles played an important role in religious thought and culture. In ancient times, Jewish and Christian scholars used them to support their faith, portraying the Sibyl as a pagan prophetess who predicted monotheism and Christianity. During the Renaissance, interest in the oracles returned, inspiring artists and writers to explore classical ideas and wisdom. Today, they are still studied for their blend of different cultural stories and their influence on prophetic and apocalyptic traditions.

The Sibylline Oracles represent the rich exchange of ideas between different cultures and religions in the ancient Mediterranean. They show how Jewish and Christian communities adapted existing pagan traditions to spread their beliefs. As a result, these writings provide valuable insight into the blending of religious ideas, the early development of Christianity, and humanity's ongoing fascination with prophecy and divine messages.

Book 1.

I will tell the story of humans from the very first to the very last. I will explain what has happened before, what is happening now, and what will happen in the future because of people's wrongdoing.

First, God wants me to explain how the world was made. Listen carefully and remember what I say, so you don't forget the commands of the most powerful King—the one who created everything by simply saying, "Let there be," and it happened.

He made the earth and placed it near the deep, dark pit below. He gave light to the world, lifted the sky high, spread the shining sea, and filled the heavens with countless bright stars. He covered the earth with plants, mixed rivers into the seas, and blended the air with gentle winds and clouds full of water. Then, he created living things—fish for the oceans, birds for the skies, wild animals for the forests, and snakes that crawl on the ground. Everything that exists now, he made with his words. It was all done quickly and perfectly because he is powerful and sees everything from above.

After creating the world, he made something special—a living being, a human, designed in his own image. He was beautiful and godlike. God placed him in a wonderful garden, giving him work to do so he could care for it. But the man, alone in this paradise, wanted someone to talk to. He prayed for a companion, someone like him.

So, God took a bone from his side and made a woman, Eve, to be his wife. She lived with him in the garden. When the man saw her, he was filled with happiness and amazement because she was a perfect match for him. They lived without shame, not even needing clothes, because their hearts were pure and free from evil.

God gave them one rule—not to eat from a certain tree. But a cunning serpent tricked them into disobeying. He convinced Eve to eat the forbidden fruit, and she gave some to the man. He listened to her and forgot God's command. Because they disobeyed, they brought suffering upon themselves instead of the good life they had.

Right away, they realized they were naked and, feeling ashamed, made clothes out of fig leaves. God was angry and sent them out of paradise to live in the world as mortals. Since they did not listen, they could no longer stay in the perfect land.

With sadness and tears, they stepped out into the world. God told them to work hard, to grow their family, and to provide for themselves through labor. From then on, survival would take effort.

As punishment, God made the serpent crawl on its belly in the dirt forever, and he put hatred between the snake and humans.

People learned to protect themselves from danger, always watching out for threats. But danger was close—both from poisonous snakes and from wicked people. Over time, the human race grew, just as God had commanded. Many generations passed, and countless people filled the earth. They built homes, created cities, and constructed strong walls with great skill. Life was long and enjoyable in those days. People didn't suffer from sickness and hardships as they do now. Instead, when their time came, they passed away peacefully, like falling into a deep sleep. These were strong, fortunate people whom God, the eternal ruler, deeply loved.

However, even they made mistakes. Foolishness led them astray. They disrespected their parents, ignored their relatives, and betrayed their own family members. Some became violent and killed others, bringing war upon themselves. Because of their evil actions, a great disaster came from heaven and wiped them out. Their lives ended, and the underworld took them in. Since Adam was the first to experience death, this place of the dead became known as Hades, and from then on, every person born on earth would eventually go there.

Even in Hades, the first people were honored because they had been the original race of mankind. But after they were gone, God created another group of people. These were clever and hardworking, focused on creating beautiful things and using their wisdom. They learned many skills—farming, carpentry, sailing, and even studying the stars. Some practiced healing with medicines, while others explored magic and different kinds of knowledge. These people were highly intelligent, always thinking and inventing, and they were strong in both body and mind.

But despite their abilities, they, too, eventually fell into darkness. Because of their actions, they were punished and thrown into a terrible place, locked away in chains. There, they suffered in an unending, fiery prison.

After them, another race appeared—fierce and arrogant. These people were cruel and constantly at war, fighting and killing one another. From them came yet another generation, the fourth race of humans. They were even worse—full of anger, violence, and wickedness. They had no respect for God or for each other, and their reckless ways led them to endless wars and bloodshed. Many were sent into the depths of the underworld because of their crimes. Others angered God so much that he cast them into the deepest, darkest pit beneath the earth, a place of eternal punishment.

But the worst was still to come. A final race of people emerged—more wicked than all those before them. They were brutal, spoke harshly, and acted without mercy. They did so much evil that they were beyond saving.

Among them, only one man stood out—Noah. He was honest, righteous, and devoted to doing good. Because of this, God spoke to him from heaven, saying, "Noah, do not be afraid. Tell the people to change their ways so they can be saved. But if they refuse to listen, I will destroy them all."

"I will send a great flood to cover the earth. Now, hurry and build a strong, waterproof house made of sturdy wooden planks. I will give you wisdom, skill, and the knowledge you need to measure and build it correctly. I will take care of everything so that you and those with you will be safe.

I am the one who exists above all things. Look inside your heart, and you will understand. I cover myself with the sky, surround myself with the sea, and the earth is beneath my feet. The air flows around me, and the stars move in harmony around me.

My name has nine letters and four syllables. The first three parts each have two letters, and the last part holds the rest. The sum of its numbers equals twice eighty plus three times thirty, plus seven. If you understand who I am, do not ignore my teachings."

When Noah heard these words, he was overcome with fear. But he understood what had to be done. He turned to the people and spoke:

"O people filled with greed and madness, do you think your wicked deeds go unnoticed? God sees everything. He has sent me to warn you so that you do not perish. Turn away from your evil ways. Stop fighting and spilling the blood of others. Do not stain the land with violence.

Respect the great and powerful Creator, who lives in the heavens. He is kind—ask him for mercy, not just for yourselves, but for the cities, the world, the animals, and the birds. Pray for his kindness.

If you do not change, the earth will be destroyed by a mighty flood. When that time comes, you will cry out in terror, but it will be too late. The sky will become wild and chaotic, and the wrath of God will come down upon you. If you refuse to turn back now, nothing will stop the punishment that is coming. Do not harm one another anymore—choose to live righteously."

But when the people heard this, they laughed and mocked Noah. They called him crazy and ignored his warning.

Noah spoke again: "You are miserable and full of evil. You have no sense of honor and only care about satisfying your own greed. You are liars, thieves, and adulterers. You speak carelessly and disrespectfully, showing no fear of God's power. Your punishment has been decided, and it will come upon you soon.

You laugh now, but you will not laugh when the great flood arrives. This world, filled with people who have turned away from God, will be wiped out in a single night. Everything—men, cities, and buildings—will be torn apart and destroyed. The entire human race will vanish. How will I mourn for you when I am shut inside my wooden house? How can I cry when the waters rise and swallow everything?

When the flood comes, the earth will float, the mountains will float, and even the sky will seem to float. Everything will be covered in water, and life as you know it will end. The winds will stop, and a new age will begin.

Phrygia, you will be the first to rise from the waters. You will be the first land where a new generation of people will grow, a fresh start for humanity."

Even after Noah warned them, the people refused to listen. Then, the Almighty appeared once more and said:

"Noah, the time has come to do everything I told you. The people have not changed. They continue to live in wickedness, just like those before them. Now, go inside with your family. Bring in the animals, birds, and creatures I have chosen to survive. I will guide them to you and make them willing to enter."

Noah obeyed. He called his family, and they all entered the wooden house. Then, the animals followed, just as God had commanded.

Noah made sure to bring everything God commanded onto the ark. Once the door was sealed tightly in place, God's plan began to unfold.

Dark clouds gathered, covering the sky and blocking out the sun, moon, and stars. The heavens grew dark as loud thunder shook the earth, and flashes of lightning lit up the sky. The winds howled, and suddenly, water poured from above in powerful torrents. At the same time, water gushed up from the ground, rising higher and higher until the entire earth was submerged.

The ark floated on the vast waters, tossed about by the raging waves. Fierce winds battered it, but it stayed afloat, cutting through the churning sea. The sound of crashing water echoed all around. As the rain poured endlessly, Noah watched and waited, following God's guidance. He had seen enough of the endless ocean and longed for solid ground again.

After many days of rain, Noah decided to check if the waters were receding. He unfastened a small opening in the ark's wall and peered outside. All he could see was water stretching in every direction. The sight of it made his heart tremble.

Then, after a long time, the rain stopped. The sky, tired of pouring water, finally began to clear. The sun appeared, its light weak and pale, shining over the flooded world. Though Noah tried to stay strong, the sight filled him with fear.

To see if land had reappeared, he sent out a dove. It flew over the endless water but found nowhere to rest, so it returned to him. The earth was still completely covered. He waited patiently and sent the dove out again after a few more days. This time, it returned with an olive branch in its beak—a sign that dry land was near. Hope filled Noah and his family, as they knew the flood was finally receding.

Then Noah released a black-winged bird, which flew off and did not return. This meant that land was close enough for creatures to survive.

The ark, after drifting for so long, finally came to rest on a narrow strip of land. It had landed on a tall mountain in Phrygia, called Ararat. This was the place where those who survived the flood would begin again. From this mountain flowed the great river Marsyas.

Once the waters had completely dried, God spoke to Noah again:

"Noah, my faithful servant, step out of the ark with your family. Bring out all the animals so they may spread across the earth once more. Multiply and fill the world. Live with justice and treat each other with kindness. A time will come when all people will be judged, but for now, rebuild the world."

Noah obeyed. He and his family stepped out onto dry land, along with the animals, birds, and every living thing that had been in the ark.

Noah, the most righteous among them, had spent forty-one days on the waters, trusting in God's plan. When he finally set foot on solid ground, a new chapter for humanity began.

This new generation, the sixth since the beginning of mankind, was golden—the best of them all. It was called the Heavenly Age because God cared for them in a special way.

I, too, was part of this new world, saved from disaster. Along with my husband, my brothers-in-law, my stepfather, my stepmother, and their wives, we endured the storm and survived the great flood.

Soon, amazing things would happen. A special flower would bloom on the fig tree, marking the beginning of a new era. Three just and noble kings would rule, each dividing the land fairly and governing with wisdom.

The earth would flourish once more, producing abundant food without effort. The people of this time would live long, free from sickness and suffering. When their time came, they would pass away peacefully, as if falling into a deep sleep. In the afterlife, they would be honored for the good lives they had lived.

The blessed ones, the fortunate heroes, were given wisdom by the Lord of Heaven. He guided them and shared his plans with them. Even in death, they would still be blessed.

But after them, another powerful race would rise—the Titans. These giants would be taller and stronger than any before them, and they would all speak the same language, just as the first humans did. However, their pride would lead to their downfall. They would become so arrogant that they would try to fight against the heavens themselves.

In response, the great ocean would flood over them, drowning the land. But despite his anger, God would hold back his full wrath, keeping his promise never to destroy the world with a flood again.

When God, the ruler of the skies, finally calms the stormy waters and sets their limits with shores and cliffs, something extraordinary will happen. A child of God will come into the world, born as a human.

His name will contain four vowels and two repeating consonants. If you add the numerical value of his letters, the total will be 888. If you understand this, you will know that he is the Christ, the Son of the eternal God.

He will not destroy God's law but will fulfill it. He will carry God's image and teach the world. Wise men will bring him gifts of gold, myrrh, and incense, and all of these things will come to pass.

One day, a voice will cry out in the wilderness, telling people to change their ways. This voice will call on everyone to purify themselves with water and turn away from evil so they may live righteously. But a cruel ruler, deceived by a dance, will order this voice to be silenced forever.

Then, a great sign will appear. A precious stone from Egypt will cause the Hebrew people to stumble, yet it will guide other nations to unity. Through him, people will come to know the one true God and find the path to salvation. He will show chosen people the way to eternal life, but those who reject him will face eternal fire.

This child of God will heal the sick and give hope to those who believe in him. The blind will see, the crippled will walk, the deaf will hear, and the mute will speak. He will cast out demons, bring the dead back to life, and walk on water. In a deserted place, he will feed five thousand people with just five loaves of bread and a single fish. Afterward, twelve baskets will be filled with the leftovers, a sign of hope for many.

But Israel will not understand. They will be blinded by their own stubbornness, refusing to listen. And when God's anger falls upon them, they will lose their faith because they rejected and killed the Son of God.

In their cruelty, they will strike him, spit on him, and offer him bitter gall and vinegar instead of food and drink. Their hearts will be filled with wickedness, unable to see the truth, trapped in darkness like blind moles or venomous snakes.

When he stretches out his hands to embrace all, they will place a crown of thorns on his head and pierce his side with a spear. As a sign of this great injustice, the sun will be darkened, and midday will turn into night for three hours. The great temple of Solomon will shake, and he will descend into the depths of the underworld, bringing the promise of resurrection to the dead.

After three days, he will rise again, showing himself to the people and teaching them. Then, he will ascend to heaven, leaving behind his teachings as a new covenant for the world.

Through him, a new people will grow—those who follow the law of the Almighty. Wise leaders will guide them, but eventually, prophecy will come to an end.

Later, when the Hebrews face the consequences of their actions, a Roman king will take away their gold and silver. Kingdoms will continue to rise and fall, bringing suffering to many. Those who become too proud and unjust will face a great downfall.

When Solomon's temple is destroyed by armored invaders, the Hebrew people will be scattered, wandering without a home. They will mix among other nations, bringing hardship upon themselves and others. The cities will suffer from violence and destruction, crying out in sorrow. And as the world turns further from righteousness, God's wrath will come upon the people for the evil they have done.

Book 2.

As I prayed, God held back my words for a time, but then he placed his divine message in my heart again. His voice flows through me, and though I don't fully understand everything I say, he guides me to speak the truth.

One day, the earth will shake with violent storms, fierce lightning, and terrifying thunder. Wild animals like jackals and wolves will grow more aggressive, and there will be terrible bloodshed. People, cattle, horses, sheep, and goats will perish. Fields will be abandoned, crops will fail, and famine will spread. Many people will be sold into slavery, and even sacred temples will be looted.

Then, a new generation will rise—the tenth race of humanity. During this time, God will put an end to idol worship and shake the mighty city of Rome, which stands on seven hills. Its vast wealth will be burned away by fire. Signs will appear in the sky, warning of the destruction to come.

The world will fall into chaos. People will turn against each other in anger, and war will rage everywhere. God will send plagues, famine, and storms to punish those who judge unfairly and act without justice. So many will die that if someone were to find another living person, they would be shocked.

But after this suffering, God will save those who are faithful. Peace and wisdom will fill the earth once more. The land will become fertile again, producing plenty of food without division or slavery. The ports and harbors will be open to everyone, and wickedness will disappear.

Then, God will show the people a great sign. A bright star will shine in the sky like a glowing crown, remaining for many days. It will be a signal for a great challenge, a test for those who seek victory. This challenge will be open to everyone, and those who succeed will be remembered forever. No one will be able to buy their way to victory with money—only those who are worthy will win.

Christ, who is pure and just, will judge each person fairly. He will reward those who stayed faithful, and he will grant an eternal prize to the martyrs who gave their lives for the truth. Those who lived righteously, remaining loyal in marriage and avoiding corruption, will receive his blessings and eternal hope.

Every soul is a gift from God, and no one should stain it with wickedness. Do not seek wealth through dishonest means, but be content with what you have. Do not take what belongs to someone else. Speak honestly and value the truth.

Do not worship false gods, but honor the one true and everlasting God. Respect your parents, fulfill your responsibilities, and avoid making unfair judgments. Do not cast out the poor or treat people differently based on appearances—if you judge unfairly, God will judge you in the same way.

Never give false testimony; always speak the truth. Stay pure and protect the love shared between people. Be fair in your dealings, and do not tip the scales dishonestly—always measure things correctly. Never swear falsely, for God despises lies.

Do not accept gifts that come from evil actions. Never steal what is meant for others, for it brings a curse across generations. Avoid sinful desires, do not spread lies, and do not take another's life.

Pay workers their wages, and do not mistreat those who are poor. Help orphans, widows, and those in need. Speak wisely and keep secrets in confidence. Never do wrong, and do not support those who act unjustly.

Give to the poor without delay—do not tell them to return another day. Share your food with those in need, and be generous, offering help with a willing heart.

Whoever gives to the poor is lending to God. Kindness and mercy will save a person when the time of judgment comes. God values mercy more than sacrifices.

Give clothes to those who have none, share your food with the hungry, and provide shelter for those without a home. Help those who are lost, guide the blind, and show compassion to those who have suffered misfortune. Life is uncertain, and no one knows what tomorrow will bring. If someone has fallen, offer them a helping hand. Defend those who have no one to stand up for them.

Everyone experiences hardships because life is always changing. Wealth is temporary, so if you have riches, use them to help others. Share what God has given you with those in need. Though all people are born into the same world, life is not always fair.

If you see someone struggling, do not mock them. Do not be harsh to those who have made mistakes. A person's true character is revealed after they are gone—whether they lived justly or not will be decided at the final judgment.

Do not dull your mind with too much wine, and avoid drinking excessively. Do not consume blood, and do not eat food that has been sacrificed to idols. Do not take up a sword unless it is for protection, and even then, it is better not to use it at all. If you kill, even in battle, your hands will be stained.

Respect your neighbor's property, and do not take what is not yours. Boundaries exist for a reason, and trespassing leads to trouble. It is good to have honest wealth, but money gained unfairly is worthless in the end. Do not harm the crops of others, and treat strangers with the same respect as your own people.

All people should treat each other with kindness because no one truly owns the land forever. Do not wish for great wealth. Instead, pray for a simple and honest life. Greed is the root of all evil. Do not crave gold or silver, as they lead people into temptation and destruction.

Money has caused wars, robberies, and even families to turn against each other. Because of it, children have betrayed their parents, and siblings have fought among themselves.

Do not deceive others or plot against your friends. Be honest in your words and actions. Do not pretend to be something you are not, changing yourself to fit every situation like a sea creature clinging to a rock. Speak truthfully and let your heart match your words.

A person who chooses to do wrong is truly evil. But if someone is forced into wrongdoing, their fate is uncertain—let each person choose what is right.

Do not be proud of your wisdom, strength, or wealth, for only God is truly wise, powerful, and rich. Do not dwell on past mistakes, as what is done cannot be undone. Control your anger, and do not act recklessly, as violence often leads to regret. Many have taken lives without meaning to, simply because they let their emotions take control.

Neither too much suffering nor too much luxury is good for people. Living in excess leads to greed and uncontrollable desires. Great wealth often leads to arrogance and cruelty. Jealousy, rage, and uncontrolled passion can drive a person to madness, leading them to act foolishly.

Righteousness is something to be proud of, while evil boldness leads to destruction. Those who are truly strong seek justice, while those who crave power bring harm. True love is good and honorable, but uncontrolled desire leads only to disgrace.

A foolish person may seem charming to others, but wisdom is found in self-control. Be moderate in eating, drinking, and conversation, for balance is the key to happiness. Too much of anything brings trouble.

Do not be envious, dishonest, or deceitful. Do not be cruel or untrustworthy. Be careful with your words and actions. Stay away from evil, and do not seek revenge—leave justice to God. Persuasion is better than conflict, for fighting only leads to more fighting. Do not trust too easily—wait to see the truth before making a decision.

This is the challenge of life, and these are the rewards. This is the path to victory and the gateway to eternal life. Those who follow it will receive honor in the eyes of God. They will pass through the gates of heaven and receive their reward.

A great sign will appear: children will be born with gray hair. People will suffer from famine, disease, and war. The times will change, and many will weep. Parents will mourn the loss of their children, burying them in the earth, which will be soaked with blood and dust.

The last generation will be full of wickedness. They will be blind to the truth, like children who refuse to understand. When women can no longer have children, it will be a sign that the end is near.

False prophets will appear, pretending to be messengers of truth. A great deceiver will rise, performing wonders to mislead the people. He will bring confusion, especially to the faithful and to the people of Israel. Many will suffer, and their belongings will be stolen. Those who hold on to their faith will face great trials, but the truth will remain.

A terrible judgment will come upon the earth. From the east, a people of twelve tribes will rise, searching for their lost Hebrew brothers and sisters who were once taken by the Assyrians. Many nations will be destroyed in this time, and these faithful and chosen Hebrews will rise to rule over others, just as they had before. Their strength will never fail.

God, who watches over all from heaven, will bring deep sleep upon many. But blessed are those who remain awake and ready when the

Master returns! Those who stay watchful, never closing their eyes, will be prepared. He will come unexpectedly—at dawn, in the evening, or at midday—but he will surely come.

When that time arrives, even the stars will be visible in the middle of the day, shining alongside the sun and moon as the final moment draws near. The prophet Elijah will descend from the heavens in his celestial chariot and show the world three terrible signs that warn of destruction.

It will be a dreadful time for pregnant women and mothers caring for newborns. Those living on the seas will face disaster. A thick, dark mist will cover the entire world, blocking out the sky in every direction. Then, a river of fire will pour down from heaven, consuming everything—land, oceans, rivers, lakes, and even the depths of the underworld. The sky itself will break apart, and the stars will fall into the seas.

People will cry out in agony as they are burned by fire and sulfur. The entire world will be reduced to ashes. Every part of creation—earth, sea, sky, light, and time itself—will be gone. No birds will fly in the sky, no creatures will swim in the sea, and no ships will sail. The land will be silent—no cattle plowing fields, no winds blowing through the air. Everything will be melted down, and only what is pure will be saved.

Then, the eternal angels of God—Arakiel, Ramiel, Uriel, Samiel, and Azael—who know every wrong deed committed by mankind, will bring all souls out of the darkness and lead them to the final judgment.

The only one who is truly eternal—the almighty God—will sit as the judge of all people. Those who have died will be restored. He will give them life again, returning their souls, voices, and bodies. Bones will be rejoined, flesh will be restored, veins and skin will be remade, and hair will grow again. They will rise as living beings once more, breathing as they did before.

Then, the great angel Uriel will break open the unbreakable gates of the underworld, tearing apart the chains of darkness. He will lead out all who have suffered in the past—the ancient Titans, the mighty giants, those who drowned in the flood, those swallowed by the sea, and those devoured by beasts and birds. Every soul, no matter how they perished, will be gathered before God's judgment.

Even those who were consumed by fire will be restored and brought before the throne of judgment. The Lord of Heaven, the one who commands the thunder, will bring an end to fate itself. He will raise the dead, sit upon his heavenly throne, and establish a mighty pillar that will never fall.

Then, Christ, who can never be corrupted, will appear among the clouds, ready to bring justice to all.

Christ will come in glory with pure angels and sit at the right hand of God on the great judgment seat. He will judge both the righteous and the wicked. Moses, the great friend of God, will return in the flesh, along with Abraham, Isaac, Jacob, Joshua, Daniel, Elijah, Habakkuk, Jonah, and those who were killed by the Hebrews.

But after the prophet Jeremiah, the Hebrews who rejected the truth will face judgment. Each person will receive the punishment or reward they deserve for what they did in life.

Everyone will pass through a river of unquenchable fire. The righteous will be saved, but the wicked will be destroyed forever. Those who committed evil—murderers, liars, thieves, homewreckers, deceivers, and betrayers—will be punished. Those who worshiped idols and abandoned the true God, those who harmed the faithful and killed the righteous, and those who pretended to be holy leaders while acting unjustly will all face judgment.

People who used their power to oppress others, those who took advantage of orphans and widows, and those who grew rich through corruption will suffer for their greed. Those who refused to care for their aging parents, broke their promises, disrespected their families, or mistreated their servants will not escape punishment.

All who abused their bodies, engaged in secret sin, encouraged abortions, or abandoned their own children will face the wrath of God. Sorcerers and those who practiced magic will be judged as well. They will be bound to a pillar surrounded by a restless fire.

God's mighty angels will chain them in burning flames and strike them with terrible punishments. They will be thrown into Gehenna, a place of endless night, where monstrous beasts roam in the darkness.

The suffering of these wicked souls will be great. A fiery wheel will circle around them as they cry out in despair. Fathers, mothers, children, and even infants will wail in agony. Their suffering will be three times worse than the evil they committed.

They will be tortured by unbearable hunger and thirst, grinding their teeth in pain. They will beg for death, but it will never come. Neither night nor rest will ease their torment. They will cry out to God, but he will turn away from them.

God had given them many chances to repent, even showing them signs through a pure virgin. But they refused to listen.

However, those who lived with kindness, justice, and faith will be led by angels through the river of fire into a place of peace. There, they will find eternal life in the presence of God.

They will come to a land where three pure streams flow—one of honey, one of wine, and one of milk. The land will be shared by all, with no fences or barriers. Food will grow freely, and no one will lack anything. There will be no more rich or poor—everyone will live together in harmony.

In that time, there will be no more rulers or servants, no rich or poor, and no kings or leaders. Everyone will be equal. No one will say, "It is night," or "Tomorrow is coming," or "Yesterday has passed." There will be no more seasons—no spring, summer, fall, or winter. There will be no more marriages, deaths, buying, or selling. The sun will no longer rise or set because God will create one endless day.

For those who have lived righteously, God will grant another gift. When they ask him for mercy, he will listen. He will rescue people from the fire and suffering, saving them from endless pain. For the sake of his faithful followers, he will remove them from their torment and bring them to an everlasting life among the immortals. They will live in the beautiful fields of paradise, near the peaceful waters of an eternal lake.

But what will become of me? I was a sinner. I wasted my life chasing meaningless things, ignoring what truly mattered. I did not honor marriage or use wisdom. Even though I lived in a wealthy home, I turned away those in need. I knowingly did what was wrong.

Yet, Lord, even though I was shameless, I beg you to save me from my suffering. I ask you to show me mercy. Let me rest now from my words, Holy One, giver of all things, King of the eternal kingdom.

Book 3.

O powerful God, who rules from the heavens and commands all things, you have placed your angels where they belong. I have spoken the truth as you instructed, but now I ask for rest because my heart is weary.

Yet, why do I still feel this urge inside me? It's as if something is pushing me forward, forcing me to speak again. I cannot remain silent—I must share what God has revealed.

People of the earth, made in God's image, why do you stray so far from the right path? Why have you forgotten the one who created you? There is only one true God, the ruler of everything, beyond human understanding. No hands have ever made him, and no artist can capture his image in gold or stone.

He alone has existed forever and always will. Who among people can look upon him with their own eyes? Who can even bear to speak his holy name? With just his word, he created everything—the skies, the seas, the shining sun, the glowing moon, and the stars that fill the night. He formed the rivers, the springs, and even the eternal fire. He set the cycle of day and night in motion.

This same God made Adam, the first human, and spread his people across the world. He created the pattern of human life and all the creatures that roam the earth, swim in the seas, and fly through the sky.

Yet, instead of honoring him, people worship false gods. They bow to snakes, make offerings to cats, and pray to idols made of stone. They sit outside temples built for gods that cannot see, hear, or move, forgetting the true Creator of all things. They foolishly believe they must protect the very God who watches over them. They place their faith in lifeless statues while ignoring the judgment of the eternal Savior who made the heavens and the earth.

What a corrupt and deceitful generation! They love violence, trickery, and lies. They have no morals, committing adultery and chasing after false gods. Greed controls them, and they take whatever they desire without guilt. The rich refuse to help the poor, and wickedness spreads like a disease.

Many women will betray their husbands, seeking love elsewhere. Marriage will lose its meaning, and faithfulness will fade away.

But when Rome gains full control over Egypt and rules over all, the eternal God will establish his greatest kingdom. A holy ruler will come to govern every nation, and his kingdom will never end.

Then, destruction will fall upon the Romans. Three great disasters will strike them, bringing chaos and ruin. Many will perish as fire rains down from the heavens, burning their homes and cities.

How terrible that day will be! When will the final judgment arrive? When will the mighty God, the King of all, bring justice to the world?

For now, cities stand tall, filled with temples, markets, grand statues, and places of entertainment where people gather. But all of this is leading them toward their downfall. A time is coming when the air will reek of burning sulfur, and people in every city will suffer greatly.

I must continue to warn all nations of the troubles that are to come.

A ruler named Beliar will rise from the Sebastenes. He will reshape the land, lifting mountains and stopping the sea. The sun and moon will seem frozen in the sky, and he will even bring the dead back to life. He will perform amazing miracles in front of the people, but his power will be built on lies. Many will believe in him, including both faithful Hebrews and those who have ignored God's teachings.

But when God's judgment comes, a massive fire will fall from the sky like a great wave crashing down. It will destroy Beliar and all who trusted in him. After this, a woman will take control of the world, and everyone will obey her rule.

This widow, now the ruler of all nations, will throw gold, silver, bronze, and iron into the sea, getting rid of the riches of mortal men. Then, the world's natural order will begin to break apart. God, who rules from heaven, will roll up the sky like a scroll, and everything in the universe will collapse.

An endless fire will pour down, burning everything—the land, the oceans, the sky, and even time itself. There will be no more day or night, no more seasons—spring, summer, fall, and winter will cease to exist.

In the middle of this final age, God's judgment will take place, bringing an end to all things.

Every land and sea, from the East to the West, will bow before the one who returns. He will take control of everything, fully revealing his divine power.

Long ago, when the people of Assyria built a great tower, God warned them of his authority. They all spoke the same language and worked together to build a tower that would reach the heavens. But God sent a powerful wind that knocked it down. He scattered the people across the world and gave them different languages. From that moment, the city became known as Babylon.

After the tower fell, people spread across the earth, creating new nations and kingdoms. The tenth generation of humans was born after the great flood. During this time, Cronos, Titan, and Iapetus became rulers. People believed they were children of the earth and sky, calling them the first kings.

The world was divided into three regions, and each ruler governed his own land in peace. They swore an oath never to fight each other, and for a time, there was balance. But when their father died of old age, the brothers broke their promises. Greed and ambition led to war, and Cronos and Titan fought for control.

The goddesses Rhea, Gaia, Aphrodite, Demeter, Hestia, and Dione stepped in and convinced the rulers to end their conflict. They agreed that Cronos, as the eldest and strongest, should rule over all.

However, Titan forced Cronos to swear that he would never have a son, ensuring that Titan would inherit the throne when Cronos grew old. To make sure this happened, the Titans watched Rhea closely whenever she gave birth. They killed every baby boy and only allowed daughters to live.

But when Rhea became pregnant for the third time, she first gave birth to a daughter, Hera. The Titans, seeing a female child, believed there was no danger and left. However, soon after, Rhea secretly gave birth to a son. She made three Cretan men swear to keep her secret and sent the baby to Phrygia, where he was hidden and raised in secret. This child was Zeus, whose name meant "sent away."

Later, Rhea protected another son, Poseidon, in the same way. Finally, she gave birth to Pluto, her third son, in Dodona. Near this place, the river Europus flowed, joining with the Peneus before emptying into the sea. The waters there became known as the Stygian River, famous for its deep and mysterious nature.

When the Titans discovered that Cronos and Rhea had secretly kept their sons alive, they gathered sixty of their strongest men. They captured Cronos and Rhea, locked them in chains, and buried them deep within the earth, keeping them under guard.

But when Cronos' sons learned what had happened, they started a great war, filling the world with chaos and destruction. This was the first war among mortals, marking the beginning of endless conflicts. God punished the Titans for their deeds, and all of them, along with the children of Cronos, perished.

As time passed, new kingdoms rose. First came the Egyptian empire, followed by the Persians, the Medes, the Ethiopians, Assyrians, and Babylonians. Then, the Macedonians gained power, later followed by another period of Egyptian rule, and finally, Rome.

Then, God placed a message in my heart, instructing me to speak of what is to come. He revealed the future of nations and their rulers.

First, the kingdom of Solomon will expand, bringing together warriors from Phoenicia, Syria, the islands, Pamphylia, Persia, Phrygia, Caria, Mysia, and Lydia, a land rich in gold.

Then, the Greeks will rise—a proud and corrupt people. After them, the Macedonians will take power, a mighty and cunning nation, sweeping across the world like a great storm of war. But God will eventually destroy them completely.

Next, a great kingdom will emerge from the western sea. This empire, strong and many-headed, will conquer vast lands and strike fear into kings. They will plunder cities and steal their treasures of gold and silver. But in time, wealth will return to the earth, and luxury will rise again.

However, these rulers will become corrupt, oppressing the people. When they give in to their arrogance and greed, their downfall will begin. Wickedness will take over, and men will turn against nature itself—men will be with men, and children will be forced into disgraceful acts.

In those days, suffering will spread everywhere. Evil will disrupt life, shattering peace and bringing destruction to all. Greed and corruption will fuel endless violence, especially in Macedonia, where deceit and hatred will grow strong.

This wickedness will continue until the seventh kingdom, when an Egyptian ruler, descended from the Greeks, will take power. But then, the people of the one true God will rise again. They will become leaders and guides, teaching others how to live righteously.

God has shown me the order of things—what will come first, what will follow, and what terrible events will mark the final days. The first to face punishment will be the Titans, for they must pay for imprisoning Cronos and his beloved wife.

After them, Greece will fall under the rule of cruel and immoral kings. These rulers will be corrupt and filled with desire, bringing endless war and suffering.

The Phrygians will be wiped out, and Troy will meet disaster. The Persians and Assyrians will face ruin, as will the people of Egypt, Libya, and Ethiopia. Trouble will spread to the Carians and Pamphylians, moving from one nation to the next, bringing hardship to all of humanity.

Why list each one separately? When the first signs come true, the rest will follow quickly. Now, I will reveal the first event that will mark the beginning of these times.

There will be great suffering for the righteous people who live near Solomon's temple—those who are descendants of just and faithful men. I will tell you their origins, their families, and their land so that you may understand.

There is a city on earth called Ur of the Chaldees. From this place comes a people known for their goodness and commitment to justice. They have always valued kindness and honorable deeds above all else.

They do not waste time studying the paths of the sun and moon or searching for signs in the depths of the ocean. They do not believe in fortune tellers, enchanters, or those who claim to see the future through meaningless tricks.

Unlike the Chaldeans, they do not rely on astrology or the stars to guide their lives. These practices only mislead people, keeping them trapped in endless searching while failing to teach them anything truly valuable.

Lies and deception led many people away from the right path, filling the world with evil. These false teachings turned people away from truth and justice. But those who live righteously reject greed, knowing that it leads to war, famine, and endless suffering. They live honestly, treating others fairly, both in cities and in the countryside. They do not steal or take what does not belong to them. They do not move land markers to claim another person's property. The wealthy do not oppress the poor or cause widows to suffer. Instead, they help those in need, sharing their wheat, wine, and oil.

Those who have more always give a portion of their harvest to those who have little, following God's law. He created the earth to be shared by all.

When the twelve tribes of Israel leave Egypt, guided by God, they will travel at night under a pillar of fire and by day with a cloud leading their way. God will choose Moses to lead them. A princess will find him as a baby near a marsh, take him in, and raise him as her own child. When he grows up, God will use him to guide his people out of Egypt to Mount Sinai.

On that mountain, God will give them his laws. He will write his commandments on two stone tablets, teaching them how to live righteously. Anyone who disobeys these laws will face punishment, either from people or from God himself. No one will escape judgment.

For seventy years, their land and temple will be left in ruins. But in the end, they will be restored and honored, just as God has promised. They must remain faithful and trust in his laws. One day, he will lift them out of their suffering and bring them back into the light.

Then, God will send a king from heaven to judge all people with fire and blood. A royal family will continue to rule, and over time, they will rebuild God's temple. The Persian kings will provide the materials—bronze, gold, and iron—because God will send them visions in their dreams. The temple will be restored to its former glory.

After I spoke these words, I prayed to God for rest. But once again, he placed a message in my heart, commanding me to share it with the world, even with the rulers of nations.

God revealed to me the suffering that would come upon Babylon for destroying his temple.

Babylon, your time of destruction is near! The earth will shake as violence spreads. The cries of destruction will rise, and God's hand will strike down your land. From above, disaster will fall upon you as punishment from heaven.

Your children will be wiped out, their souls taken by the Eternal One. It will be as if you had never existed. But your land will be filled with blood, just as you once shed the blood of good and righteous people. Their cries still reach the heavens.

Egypt, you too will suffer. You believed you were safe, but disaster will strike. A sword will pass through your cities, bringing death, famine, and ruin. This suffering will last through seven generations of kings before it ends.

Land of Gog and Magog, near the rivers of Ethiopia, your streets will be covered in blood. Your land will become a place of judgment, filled with death and destruction.

Libya, your suffering is coming as well. The western lands and the sea itself will face a dark day. War will chase you down, bringing violence and destruction.

You destroyed the house of the Immortal God, tearing it apart with iron. Because of this, your land will be covered with the dead—killed by war, famine, disease, and enemy attacks. Your cities will be abandoned, left in ruins.

In the West, a bright star will appear in the sky, a comet bringing a warning. It will be a sign of war, famine, death, and the downfall of kings and rulers.

Strange signs will appear everywhere. The Tanais River will break away from Lake Maeotis, and new land will rise where there was once only water. Rivers will change their course, and deep cracks will open in the earth, swallowing entire cities.

Many places will be destroyed, including cities in Asia—like Iassus, Cebren, Pandonia, Colophon, Ephesus, Nicaea, Antioch, Sinope, Smyrna, Myrina, and Gaza. In Europe, places like Tanagra, Clitor, Basilis, Meropeia, Antigone, Magnesia, Mycenae, and Oiantheia will also fall.

Egypt's power will soon come to an end. The past will be better than the future for the people of Alexandria.

Rome will be repaid for its actions. The riches it once took from Asia will be returned threefold. The destruction it caused will be repaid with its own suffering.

Just as many people from Asia were once taken as servants in Rome, even more Romans will become slaves in Asia. They will be trapped in poverty, buried in debt, and stripped of their wealth.

Rome, you were once proud, celebrating in luxury, drinking at grand feasts. But now, you will be a slave, forced into shameful marriages. The women of Rome will be humiliated, their beauty taken away. They will be dishonored, their hair cut, their dignity lost.

The rulers of the land will be overthrown, and no one will show them mercy. Corrupt leaders will destroy the people, using their power for evil instead of justice.

Samos will turn to dust, Delos will lose its brightness, and Rome will become nothing but an empty space. But everything God has planned will happen exactly as he said. Peace will come to the lands of Asia, and Europe will finally know happiness. The land will be full of life, free from storms and disasters. The air will be clean, and the fields will be filled with food. Animals, birds, and all living things will flourish.

People will live in joy, their homes filled with happiness. From the heavens, justice and order will guide humanity. People will treat each other with kindness and trust, welcoming strangers with open hearts. Evil will disappear—greed, jealousy, anger, and foolishness will be gone. There will be no more poverty, violence, war, or crime.

But Macedonia will bring pain to Asia, and Europe will suffer from a corrupt and undeserving family. Born from both rulers and slaves, they will conquer Babylon and claim to rule all the lands under the sun. But in time, they will fall, and their name will only be remembered far into the future.

Then, a man unlike any before will rise in Asia. He will wear a purple robe, be ruthless and unfair, and crave power. The god of thunder will push him forward. Under his rule, Asia will suffer, its lands soaked in blood. But the underworld will take him away, and the very people he tried to destroy will wipe out his family.

Out of this chaos, a new leader will rise. He will cut down a powerful ruler and take his place. A son will kill his own father, a great warrior, and the god of war himself will fall at the hands of his grandson. Soon after, the new ruler will take control, replacing the old one.

In Phrygia, a terrible sign will appear. When the cursed descendants of Rhea suddenly vanish overnight—along with an entire city—it will be a warning of disaster. The god of the sea, Poseidon, will shake the land, tearing down walls and breaking the earth open. The city of Dorylaeum in Phrygia will be destroyed, and the world will forever remember this as the time of the "Earth-shaker."

But this event will not bring peace—it will mark the beginning of great suffering. A terrible war will spread across the land. The descendants of Aeneas, born from the bloodline of Ilus, will bring destruction, and their city will eventually fall to greedy invaders.

Troy, your fate is sealed. In Sparta, a beautiful and powerful woman will rise, and her actions will send waves of destruction across Asia and Europe. But you, Troy, will suffer the most. Endless war, pain, and sorrow will follow, yet your name will live on in history.

During this time, an old man will appear. He will call himself a writer, though no one will know where he truly comes from. His eyesight will fail, but his mind will stay sharp, and his poetry will be admired. He will claim to be from Chios and write about Troy, but his words will not be entirely true—only beautifully written.

This man will use my words, opening my writings for the first time. But he will change them, turning them into grand tales of warriors in shining armor. He will make Hector, son of Priam, and Achilles, son of Peleus, into legendary figures of war. He will add gods to their battles, creating false stories. His version of history will make the deaths of these warriors seem noble. But in return, he will receive great praise for his work.

A group from Lycia will bring suffering to the people of Locri. Chalcedon, which guards a narrow sea passage, will one day be attacked by a young warrior from Aetolia. The wealthy city of Cyzicus will be shattered by the sea. Byzantium, the city of Ares, will be destroyed by an army from Asia, drowning in blood and sorrow.

The great mountain Cragus in Lycia will shake, causing deep cracks to form. Water will rush from the broken rock, and the oracles of Patara will fall silent forever. Cyzicus, near the wine-rich Propontis, will be hit by the raging waves of the Rhyndacus River.

Rhodes, island of the sun, you will enjoy many years of freedom and success. You will rule the seas and grow strong. But in time, greedy men will take over, weighing you down with heavy burdens because of your beauty and wealth.

A massive earthquake will strike Lydia, shaking Persia and causing terrible destruction. The people of Europe and Asia will suffer greatly.

A cruel king from Sidon will bring destruction to the seafaring people of Samos. Blood will flow through the land and mix with the sea. The noble women and brides will cry in sorrow, grieving for their husbands and sons who have been lost.

Cyprus, a powerful earthquake will shake your land, taking many lives. The underworld will claim the souls of those who die together.

Trallis, near Ephesus, despite your strong walls and great wealth, you will also fall to an earthquake. The ground will split open, releasing boiling water from deep below, and the earth will tremble with unstoppable force.

Flames and the choking smell of burning sulfur will destroy those who are trapped by its power. One day, Samos will build great royal palaces.

Italy, you will not be invaded by enemies from other lands, but your own people will turn against each other in violent battles. Bloodshed between your own tribes will leave you empty and broken. In the end, you will be covered in ashes, never realizing that your downfall was your own doing. You will no longer be a land of great leaders, but a place where wild beasts roam.

A merciless ruler will rise from Italy and bring destruction. Laodicea, the beautiful city of the Carians by the Lycus River, you will fall, mourning in silence for the ancestors you once honored.

The Thracian Crobyzi will rise from the Haemus Mountains. The people of Campania will suffer, grinding their teeth in despair as famine devours them. Corsica will grieve the loss of its elders, and Sardinia will sink beneath the sea under the force of powerful storms and divine punishment. Those who live by the sea will look on in shock at this great disaster.

So many young women will die. Countless young men will drown in the deep, their bodies lost without a proper burial. Helpless children will perish, and great treasures will be swallowed by the sea.

Mysia will see a royal family rise to power. Chalcedon will soon come to an end. Galatia will suffer greatly, and Tenedos will face its final and most terrible disaster. Sicyon and Corinth will cry out in sorrow, yet they will still boast, as flutes play their mournful songs.

When my soul had a moment of rest, God placed another message in my heart. He commanded me to reveal what is yet to come.

A terrible fate awaits the people of Phoenicia—the men, the women, and all the cities by the sea. Not one of you will survive to see the sunlight, for your time is coming to an end. Your people will vanish because of your lies and wicked ways. You spoke false words, lived without honor, and turned your backs on God, the true King.

Because of this, destruction will come upon you. A great fire will rise from the earth, consuming your cities until only ruins remain where mighty buildings once stood.

Crete, a disaster is coming. A devastating blow will strike, and the Almighty will bring great destruction. Smoke will rise, covering the sky. The flames will not die out, and your land will burn until nothing is left.

Thrace, your people will be enslaved. The Galatians will join forces with the sons of Dardanus, invading Greece and bringing devastation. Your land will be taken from you, and you will give away much but receive nothing in return.

Gog, Magog, and all the people of Mardia and Daia—great suffering is waiting for you.

Lycia, Mysia, and Phrygia will also feel the weight of this disaster. Many people from Pamphylia, Lydia, Caria, Cappadocia, Ethiopia, and Arabia will fall. There are too many to name, but every nation will be struck by the plague that the Almighty will send.

A brutal army will invade Greece. They will kill strong leaders and slaughter countless sheep, horses, mules, and cattle. They will burn down homes, leaving nothing behind. They will take many prisoners—men, women, and children—forcing them into slavery. Young brides will be torn from their homes, their feet barely touching the ground as they are dragged away. Their captors, speaking a language they do not understand, will bind them in chains and treat them cruelly.

No one will come to save them. They will watch as their enemies take everything—their wealth, their homes, their lands. Their legs will tremble in fear.

A hundred men will flee, but one enemy will chase them all down. Five soldiers will be enough to scatter an entire army. The people will turn against one another in shameful battles, bringing joy to their enemies but suffering upon themselves.

Then, all of Greece will be enslaved. War and disease will come together, bringing suffering to everyone. God will turn the sky into hard bronze and the earth into iron. No rain will fall, and drought will cover the land. Crops will wither, and the ground will crack under the burning heat.

People will cry out in pain and hunger, but there will be no relief. The Creator of heaven and earth will send fire down, and only one-third of mankind will survive.

Greece, why do you put your trust in leaders who are only human and cannot escape death? Why do you make foolish offerings to idols? Who convinced you to follow this false path and turn away from the one true God?

Remember the Almighty and do not forget his name. It has been 1,500 years since proud rulers first led Greece. They introduced false gods to the people, creating statues for worship and filling hearts with foolish beliefs.

But when the wrath of the Almighty comes, you will finally see the truth. People will lift their hands to the sky in despair, crying out for help. They will beg for a savior to rescue them from the great disaster that is coming.

Listen and remember these words in your hearts, for hard times will come in the years ahead.

If Greece offers sacrifices of bulls and cows to the great God, she will escape from war, fear, and deadly sickness. She will break free from oppression. But until that time, there will still be people who turn away from God, even when the final day arrives.

You must not make offerings to God until everything he has planned is completed. Nothing he has decided will fail to happen, and events will unfold with great force.

One day, there will be a new generation of faithful people who follow God's teachings and live by his wisdom. They will honor his temple with offerings, burnt sacrifices, and gifts of bulls, rams, lambs, and the best of their flocks. They will offer these sacrifices with pure hearts on God's great altar.

Living by his laws, they will be blessed, and their cities and fields will be full of riches. God will send prophets to guide them, bringing great joy to all people. Only they will receive his wisdom and have faith and righteous hearts.

They will not be misled by worthless things or worship idols made of gold, bronze, silver, or stone. They will not bow to statues or lifeless images created by human hands. Instead, they will raise their hands to heaven in prayer, waking early each morning to cleanse themselves with water and honor the eternal God.

They will respect their parents and remain faithful to their marriages. They will not engage in disgraceful acts like the Phoenicians, Latins, Egyptians, Greeks, Persians, Galatians, and many others who have broken God's laws. Because these nations have disobeyed him, they will suffer. God will send famine, pain, war, disease, and sorrow upon them for refusing to worship him.

Instead of honoring the one true God, people will worship man-made idols. But when a young king—Egypt's seventh ruler since the Greeks took control—comes to power, they will realize their mistakes.

A great ruler from Asia will come like a fiery eagle, bringing his army on foot and horseback. He will spread destruction across the land, fill it with suffering, and overthrow Egypt. He will take away its wealth, carrying it across the sea.

Then, before the eternal King, people will kneel in submission, bowing on the earth that has long provided for them. All the idols they made with their own hands will be destroyed by fire.

At last, God will bless the people. The land will flourish, and trees and flocks will provide in abundance. Wine, honey, milk, and wheat—the best gifts for humankind—will be plentiful.

But you, who live with clever tricks and deceit, do not waste time. Stop hesitating and turn back to God. Offer sacrifices of bulls, lambs, and goats as the seasons change. Seek his mercy, for he is the only true God—there is no other.

Live with justice and do not oppress others. This is what the Almighty commands of all people.

Pay attention, for the wrath of God will be great. A terrible plague will sweep across the earth, and people will face harsh judgment. Kings will rise against kings, stealing lands and waging war. Nations will destroy each other, and rulers will rob entire tribes of their homes and riches.

Leaders will flee to other lands, and new rulers will take control. Foreign powers will invade Greece, taking its wealth and leaving the land in ruins. The people will turn against each other, fighting over gold and silver. Greed will lead their cities like a heartless ruler.

Many will die without a proper burial, their bodies left for vultures and wild animals. The earth will be covered in their remains, left unplowed and abandoned for years. The land will show the shame of those who turned away from what is right, with scattered weapons and broken shields. Even the forests will be untouched, as no one will cut wood for fire.

Then, God will send a king from the East to bring peace to the world. He will put an end to war, killing some enemies and forcing others to swear oaths of loyalty. But he will not act on his own, instead following the commands of the Almighty.

Under his rule, the temple of God will be restored, filled with treasures of gold, silver, and fine cloth. The land and sea will once again overflow with abundance.

But jealousy will turn kings against one another, leading them to fight and destroy themselves. In their greed, they will seek to plunder God's temple and harm the most righteous people. When they arrive in the land, wicked rulers will place their thrones around the holy city, bringing people who do not follow God's ways.

Then, the Almighty will speak with a mighty voice, calling out to those who are foolish and arrogant. His judgment will come upon them, and they will be destroyed by his power.

Flaming swords will fall from the sky, and bright lights will blaze down upon the earth. The land will shake, and the creatures of the sea and land, along with all people, will tremble before the Immortal One. Fear will spread everywhere.

Towering mountains will crack apart, and great hills will crumble. Deep chasms will open, revealing the dark underworld. Valleys will be filled with the dead, and rivers will run red with blood, spilling into the plains.

The strongholds of the wicked will collapse because they ignored the laws and judgment of the Almighty. With reckless hearts, they attacked the temple, raising their weapons against it. But God will judge them with war, fire, storms, and destruction. Fiery hail, sulfur, and great stones will rain from the sky, and even the animals will suffer.

Then, they will finally recognize the true God. But it will be too late. Cries of terror will rise from the earth as countless people perish. The wicked will be drenched in blood, and the land itself will drink deeply of the fallen. Wild beasts will feast on the dead.

All these things have been revealed to me by the eternal God, and they will surely come to pass. Everything he has placed in my heart is true, for his spirit never lies.

But the children of God will return to live in peace around the temple. They will rejoice in the blessings given by their Creator, the just Judge and mighty King. He will be their protector, surrounding them like a wall of fire. They will live in their cities and land without fear of war, for they will no longer need weapons to defend themselves.

There will be no more battles, for the Almighty himself will guard them and keep them safe.

Then everyone will see how much God loves those who stay faithful to him. Even the sun, moon, and sky will help them in their struggles and lead them to victory.

In that time, people will sing songs of praise:

"Let us fall to the ground and pray to the eternal King, the one true God. Let us go to his temple together, for he alone is Lord. Let us follow the laws of the Most High, for they are the most just on earth. We have wandered far from his path, foolishly worshiping idols made by human hands, statues of dead men."

The hearts of the faithful will be moved, and they will cry out:

"Come, let us bow before God's house and sing to him with joy. Our land has been freed from enemies, and for seven generations, we will be surrounded by their abandoned weapons—shields, helmets, bows, and arrows. No one will need to cut wood for fire, for we will have all we need."

But Greece, you must let go of your pride and seek wisdom. Turn to the Almighty with humility, for he is merciful but also just.

Do not stir up conflict with those who come from God's holy land. Do not try to move what should remain still, or awaken a beast from its den, or invite disaster upon yourself. Let go of your arrogance and seek peace. Serve the Almighty so that you may share in his blessings.

When the final day comes, and God's judgment arrives, the earth will produce endless fruit—wheat, wine, and oil in abundance. Sweet honey will pour from the heavens, and trees will be heavy with fruit. Sheep and cattle will multiply, and the fields will be rich with crops. Rivers of milk will flow freely, bringing nourishment to all.

There will be no more war, no more suffering. The earth will no longer shake in fear. Drought, famine, and storms will disappear. A great and lasting peace will cover the world. Kings will no longer fight but will live as friends until the end of time.

God, who rules over the stars, will bring justice to all people, judging their actions. He alone is God, and there is no other. He will put an end to human cruelty, burning away evil with fire.

So change your ways. Turn away from false worship. Serve the one true God. Stay away from adultery and wickedness. Raise your children and do not take innocent lives, for God is angered by those who commit such sins.

Then, God will establish a kingdom that will last forever. He will fulfill his promises to the faithful, opening every land and blessing them with eternal joy. People from every nation will bring offerings of incense and gifts to his temple, for there will be no other place of worship—only the one God has chosen for his people.

In those days, travel will be safe. Roads through valleys, mountains, and even across the sea will be easy to cross. Peace will cover the land. Prophets of God will remove all weapons, for they will be the true leaders and just rulers of the people. There will be no more greed, only fairness and righteousness, for this is God's will.

Rejoice, for the Creator of heaven and earth has given you joy. He will dwell among you, and all life will live in harmony. Wolves and lambs will graze together. Leopards will rest beside baby goats. Bears and cattle will share the same pastures. Lions will eat straw like oxen, and little children will lead them without fear.

Deadly snakes and scorpions will no longer be a danger. Infants will sleep beside them, unharmed, for God's hand will protect them all.

Now, I give you this clear sign so that you will know when the end of all things is near.

When the end comes, there will be signs in the sky. At night, swords of light will stretch across the heavens, pointing both east and west. A thick cloud of dust will rise and spread across the earth.

The sun will darken in the middle of the sky, and the moon's light will disappear and then return again. Blood will drip from the rocks as a warning, and in the clouds, people will see visions of armies—soldiers on foot and horseback, moving like hunters chasing wild animals through heavy fog.

This will be the final moment, the time when God, who rules from the heavens, brings everything to an end. But before that happens, all people must turn to him and offer their devotion to the one true King.

I tell you these things because I left the great city of Babylon behind to warn the people of Greece about God's coming wrath. Fire will be sent down from above as punishment.

I was given the gift of prophecy to reveal divine mysteries to mortals. Some will say that I come from a faraway land, that I was born in Erythrae, and call me a liar. Others will claim I am a Sibyl, the daughter of Circe and Gnostos, mad and deceitful. But when these events come to pass, you will remember my words. No one will doubt that I was a prophet of the great God.

He showed me what happened to those who came before us and revealed the first moments of creation. He placed knowledge in my heart so that I could speak of the future and remind people of the past.

Long ago, when the earth was covered in a great flood, only one righteous man was saved. He and his family, along with animals of every kind, sailed in a wooden ark so that life could begin again.

I was married to his son and came from his bloodline. Because of this, I have seen both the first and the last of all things, and I have spoken the truth as it was revealed to me.

Book 4.

People of Asia and Europe, listen carefully to my words. I speak the truth, not as a false prophet of Apollo, whom foolish men call a god, but as a messenger of the one true God. He is not a lifeless statue made by human hands, nor is he confined to a temple built of stone.

No one can see or measure him, for he is beyond human sight. He rules over both night and day, the sun, moon, and stars, the land, the seas, the rivers, and all living creatures. He controls the rain, which brings fruit to the trees, grain to the fields, and oil from the earth.

God has placed this message in my heart, urging me to tell people what has happened and what will come, from the first generation to the eleventh. Everything I say will be proven true by the events that unfold. Listen carefully to the Sibyl, who speaks only what is right.

Blessed are those who love and honor the mighty God. They should praise him before they eat or drink and live with faith. They must not worship false idols, lifeless statues made of stone. Those who choose to live in wickedness will mock the righteous, accusing them of wrongdoing while they themselves commit evil.

People are slow to believe the truth, but when the day of judgment comes, when God separates the righteous from the wicked, then they will understand. The sinful will be cast into darkness, while the faithful will remain on the earth, blessed with life and grace.

This will happen in the tenth generation. But first, I will tell of the events from the beginning.

The Assyrians will be the first to rule over mankind, holding power for six generations after God, in his anger, covered the world with a great flood. Then the Medes will rise and take control, but they will rule for only two generations.

During this time, the world will see terrifying signs. At midday, darkness will cover the sky, and the sun, moon, and stars will vanish. A great earthquake will shake the land, destroying many cities. Even the islands of the sea will be revealed as the waters shift.

When the Euphrates River turns red with blood, a great war will break out between the Medes and the Persians. The Medes will suffer a crushing defeat, forced to flee beyond the Tigris River. The Persians will then rise to power, becoming the strongest nation in the world for a generation.

But their rule will bring chaos. War, murder, rebellion, and destruction will spread. Towers will crumble, cities will fall, and many will be exiled. Greece, in its great strength, will cross the Hellespont, bringing sorrow to Phrygia and disaster to Asia.

Egypt, the land of many fields, will suffer greatly. A terrible famine will strike, and for twenty years, the land will be barren. The Nile, the river that nourishes the crops, will no longer provide for the people.

Deep underground, dark waters will continue to flow.

A powerful king will come from Asia, leading countless ships and carrying a spear. He will travel across the sea, cutting through mountains to clear his path. After fleeing from battle, he will seek safety in Asia.

In Sicily, a fiery river will set the land on fire as Mount Etna erupts, sending flames into the sky. The great city of Croton will collapse into a deep pit.

In Greece, brutal wars will break out. Cities will be destroyed, and many lives will be lost. Both sides will suffer equally.

When the tenth generation of people arrives, Persia will fall under oppression and fear. The Macedonians will rise to power and destroy Thebes. The Carians will take over Tyre, wiping out its people.

Babylon may look strong, but it will prove weak in battle. Its high walls will not protect it as expected. Macedonians will settle in Bactria, while people from Susa and Bactria will escape to Greece.

One day, the Pyramus River will flood and reach the sacred island. When the earth shakes, cities like Cibyra and Cyzicus will fall apart. Sand will bury Samos, erasing it from sight. Delos will disappear completely. Rhodes will suffer its worst disaster yet.

The Macedonian empire will not last. A great war will come from the west, and Italy will take control. The world will be ruled by its heavy hand, and even the Italians themselves will suffer under its power.

Corinth, your downfall is coming. Carthage, your towers will crumble, and your walls will be flattened to the ground.

Laodicea, an earthquake will tear you down, turning everything to rubble. But in time, you will be rebuilt.

Lycia and Myra, the earth will never hold you steady. You will fall, and like strangers in your own land, you will beg for escape. Patara will be silenced by storms and earthquakes, punished for its wickedness.

Armenia, you will become enslaved.

War will come from Italy to Jerusalem, destroying God's great temple.

When the people of Jerusalem turn away from their faith and commit terrible crimes near the temple, a powerful ruler will rise in Italy. Like a fugitive, he will flee across the Euphrates, staying hidden from sight. He will commit terrible crimes, even killing his own mother, believing in his own cruelty.

Blood will flow in the streets of Rome as many fight for the throne. The land will be soaked in the blood of its own people.

A leader from Syria will rise in Rome. He will burn the temple in Jerusalem and massacre many Jewish people, leaving their homeland in ruins.

A massive earthquake will destroy Salamis and Paphos. Huge waves will crash over Cyprus, drowning the island.

When fire erupts from deep within Italy, shooting flames into the sky, many cities will burn, and countless lives will be lost. Thick black ash will cover the land, and red dust will rain down from the heavens.

This will be a sign of God's anger, because the people will have destroyed those who were faithful to Him.

War will break out in the West, and chaos will spread. A fugitive from Rome will rise, carrying a great spear, bringing destruction as he marches forward.

The mighty Euphrates River will be filled with countless bodies.

Oh, unfortunate Antioch, you will no longer be called a city when you fall to enemy spears because of your own mistakes.

On the island of Scyros, a deadly plague and violent battles will bring destruction.

Oh, poor Cyprus, a massive wave will rise and cover your land, tossing you into chaos as fierce storm winds whip through your shores.

Wealth will flow into Asia, the same riches that Rome once stole and stored in its grand homes. But Rome will be forced to return twice as much, and with it, war will spread even more.

The beautiful cities along the Maeander River, surrounded by tall towers, will be ruined by a terrible famine when the river hides its dark waters.

When humanity turns away from righteousness, and truth and justice disappear from the world, people will become reckless and violent. They will take pleasure in doing wrong, ignoring the good, and destroying the righteous. Their hands will be stained with blood, and they will celebrate their own cruelty like foolish children.

When this happens, know that God will no longer be patient. His fury will be unleashed, and he will destroy the human race with a great fire.

Oh, foolish mortals, change your ways before it is too late! Do not provoke the mighty God's wrath. Put away your swords and stop your violence. End the killing, the cruelty, and the senseless destruction. Wash yourselves in pure waters, raise your hands to heaven, and ask for forgiveness. If you truly repent and turn back to what is right, God will have mercy on you. He will hold back his anger and spare the world.

But if you refuse to listen and continue in your wicked ways, a great fire will cover the earth. A terrifying sign will come, with the sound of trumpets and clashing swords at sunrise. The whole world will shake from the noise. God will burn the land, destroy every city, and wipe out the rivers and seas. Everything will turn to black ash.

When everything has turned to dust and the flames have died down, God will calm the great fire he has sent. Then, from the ashes, he will rebuild the bones of men and bring them back to life, just as they were before.

At that time, he will hold the final judgment. God himself will judge the world once more. Those who lived with wicked hearts will be buried beneath the earth, never to rise again.

Book 5.

Listen carefully and remember the difficult times that will come for the people of Latium.

First, after the rulers of Egypt have fallen and been buried, and after the man from Pella who conquered both the East and West has died—his body abandoned by Babylon, proving false the claim that he was a son of Zeus—a new ruler will rise. He will come from the bloodline of Assaracus, a descendant of Troy, and he will survive the fires of destruction. After him, many leaders will take power—some will be great warriors, while others will be inexperienced and weak.

The first of these rulers will have a name connected to the number twenty. He will be strong in war. After him, another will rule, with a name that begins with the first letter of the alphabet. He will bring

Thrace and Sicily under his control, and Memphis will fall because of weak rulers and a free-spirited woman who is lost at sea.

This ruler will establish laws and rule over many nations. After a long time, he will pass his power to another whose name is linked to the number three hundred and a river. This new ruler will command Persia and Babylon and will defeat the Medes in battle.

Then another ruler will come, marked by the number three. After him, a new leader will rise, his name connected to the number twenty. He will sail far across the ocean and reach the shores of Italy.

Next, a ruler with a name linked to the number fifty will appear. He will be like a fierce serpent, bringing war and destruction. But eventually, he will turn on his own people, causing chaos. He will seek fame through chariot races and great battles, leaving behind a trail of blood. He will carve a path through two seas, but in the end, he will disappear. He will return, claiming to be a god, but the true God will show his power over him.

After this, three kings will rise, only to destroy each other. Then, a cruel ruler whose name carries the number seventy will come to power. He will bring great suffering to the faithful. His son, marked by the number three hundred, will take the throne next. After him, a ruler with the number four will rise, bringing destruction.

Another leader with the number fifty in his name will come next. Then, one marked by the number three hundred will take the throne—a warrior from the Celtic lands. But he will not escape a terrible fate. After many battles, he will die on foreign land, buried in dust named after the Nemean flower.

Following him, another ruler will rise, wearing a silver helmet and carrying the name of a sea. He will be the greatest of them all, wise in his ways. His rule will bring change, and his descendants will witness these events unfold.

After him, three more rulers will take power, but the last one will rule for a long time.

I am tired, weighed down by this vision, but I must share what I have seen.

First, wild women will dance around the steps of your once-glorious temple, bringing destruction. A time will come when the Nile overflows, covering Egypt with sixteen cubits of water. It will flood the land, bringing rich soil for crops, but it will also mark a time of great change. Egypt's beauty will remain, but its power will fade.

Memphis, you will mourn more than any other city in Egypt. Once strong, you will become poor and weak. The great voice from the heavens will cry out:

"Oh, mighty Memphis, you once ruled over fearful men, but now you will suffer! You will cry out in pain and sorrow. Only then will you recognize the eternal God, the one who rules above the clouds.

Where is your pride now? You stood against my chosen people and brought suffering upon the righteous. Now you will pay the same price. You will no longer stand among the blessed."

"You have fallen from the stars and will never rise to heaven again."

God commanded me to deliver this final warning to Egypt, for the time will come when people become completely corrupt. Evil men will continue their wicked ways, but punishment awaits them—the wrath of the mighty one in heaven. Instead of worshiping God, they will bow to stones and animals. They will fear lifeless things that have no voice, no mind, and no power to hear their prayers.

It is not right for me to name these false gods, but they are nothing more than idols, created by human hands. People have made their own gods from wood, stone, bronze, gold, and silver, melting them in fire and trusting in things that cannot think or speak.

Thmois and Xois will suffer greatly, and the great temples of Heracles, Zeus, and Hermes will be struck down.

Alexandria, once a powerful city, will face endless war and disease. Because of your pride, you will suffer the same pain you once caused. Silence will fall over you for many years, and your time of restoration will be distant. No longer will you enjoy your rich and abundant pleasures.

A Persian invader will come into your land like a storm, bringing destruction. He will fill the land with bloodshed, slaughtering many. A ruthless, mindless leader will rush into battle with countless soldiers, leaving nothing but ruin behind.

Wealthy cities will become exhausted and broken. All of Asia will weep, once adorned with riches, now stripped of everything she once loved. The conqueror who seizes Persia will bring war, wiping out almost all life. Only a third of the people will survive his wrath.

From the West, he will move swiftly, taking control of lands, destroying everything in his path. When he reaches his full power and is feared by many, he will set his sights on attacking the city of the blessed.

But God will send a mighty king against him—one who will destroy the strongest rulers and bravest warriors. This will be the moment when the immortal one brings judgment upon the world.

Oh, my sorrowful heart! Why must I speak of these painful times? Egypt's rule over many lands will bring suffering.

Turn instead to the East, to the Persians—people lacking wisdom—and warn them of what is happening now and what is still to come.

The Euphrates River will overflow, bringing destruction. The flood will wipe out the Persians, Iberians, Babylonians, and the war-loving Massagetae, who foolishly rely on their bows. Fire will spread across all of Asia, burning so brightly it will be seen from the islands.

The once-revered city of Pergamos will be completely destroyed. Pitane will become a wasteland.

All of Lesbos will sink beneath the waves, disappearing into the sea. Smyrna, once honored, will be thrown from her cliffs, crying out as she perishes completely.

The people of Bithynia will mourn over their land, reduced to ashes. Great Syria and the many tribes of Phoenicia will also fall into ruin.

Lycia, so much suffering will come upon you! The sea itself will rise against you, flooding the land and bringing disasters.

A great earthquake will strike, and bitter waters will crash onto the once-fragrant lands of Lycia, turning beauty into devastation.

Phrygia will face terrible destruction, bringing sorrow to Rhea, the mother of Zeus, who once stayed there for a long time.

The sea will rise and wipe out the Centaur race and a fierce nation, while the land of the Lapiths will be swallowed beneath the earth.

The deep and fast-flowing Peneus River will flood Thessaly, sweeping away its people. Eridanus, once known for its strange waters, will bring further disaster.

Greece will suffer terribly, and poets will mourn for her when a ruler from Italy strikes at the isthmus. This powerful Roman king, who is treated like a god, will rise to power. Some will even claim he is the son of Zeus and Hera.

This ruler, who charms people with his beautiful voice and songs, will commit great evil. He will kill many, including his own mother.

A feared and shameless leader from Babylon will flee. People will despise him because of his many crimes—he took the lives of countless people, defiled the wombs of women, and committed terrible sins against his own wives.

He will seek refuge with the kings of the Medes and Persians, the same people he once honored and helped rise to power. But alongside wicked men, he will secretly plot against an unwanted nation.

He will seize God's holy temple and attack its people. Those who entered the temple in faith will be burned. When this man appears, the whole world will tremble. Kings will fall, but his power will remain, leading to the destruction of a great city and the suffering of righteous people.

In the fourth year of these events, a great star will shine brightly, dominating the sky. It will hold special significance, tied to the honor once given to Poseidon.

Then, another massive star will fall from the heavens into the sea, setting the waters ablaze. Its fire will reach Babylon, Italy, and other lands, bringing destruction as punishment for the deaths of many faithful Hebrews and innocent people.

You, sinful city, will suffer greatly. You will be abandoned, left empty and desolate for ages. You will despise your own land, but it will be too late.

You gave yourself to sorcery, committed adultery, and indulged in shameful and unnatural acts. You were an evil, corrupt, and unjust city—more cursed than all others.

Oh, city of Latium, you are impure in every way! Like a frenzied woman who delights in snakes, you will sit alone, widowed, on the banks of the Tiber River.

The river itself will mourn for you, its once-proud partner, now filled with blood and sin.

Did you not understand the power of God and his plans? You thought yourself untouchable, saying, "I am invincible—no one can destroy me."

But now, the eternal God will bring ruin upon you and everything you possess. Your banners will no longer wave in the land as they once did when you received honor from God.

Now you will be left alone, abandoned in a fiery prison, dwelling in the burning depths of the underworld.

And once again, I grieve for Egypt, lost in blindness.

Memphis, you will be overrun with the dead. The pyramids will echo with the cries of those who suffer.

Python, once known as the twin city, you will fall silent for generations, forced to abandon your wickedness.

A land filled with suffering and pain, a place of endless sorrow—you will be left as a weeping widow. You ruled the world for many years, but your time is over.

When the unclean land is covered by the white robes of purity, I will wish I had never been born to witness it.

Thebes, where is your great strength now? A fierce man will slaughter your people. You will be left alone, dressed in mourning, crying out in despair.

You will be punished for the sins of your past. Those who remain will see the price of lawlessness and evil.

A mighty leader from Ethiopia will rise and overthrow Syene, showing the power of his people.

Dark-skinned warriors from India will take over Teucheira.

Pentapolis, a mighty leader will burn your cities to the ground. Libya, full of sorrow, who will explain your mistakes? And Cyrene, who will mourn for you? Even in your final moments, your cries will not stop.

Across Britain and Gaul, where gold is plentiful, the ocean will roar with the sound of battle, stained red with blood. These lands brought suffering upon God's people when a king of the Sidonians, a Phoenician leader, led a great army of Gauls from Syria.

This army will bring destruction upon you, Ravenna, leading you to slaughter.

People of India and Ethiopia, beware. When Capricorn and Taurus align in the sky, with Virgo rising, the sun will lead the heavens in a new direction. A massive fire will descend from the sky, reshaping the stars in battle, and Ethiopia will be consumed by flames, drowning in sorrow.

Corinth, prepare to weep for your own destruction. When the three sisters of fate spin their threads, they will guide a deceiver to power. He will stand before the isthmus, where a man once carved through rock with metal tools. This leader will bring ruin to your land, as fate has decided.

God will grant him the power to do what no other ruler before him could achieve. He will first cut down three great leaders, feeding their remains to others. Unholy kings will feast upon the flesh of their own families. Death and terror will spread across the land because of a great city and a righteous people who were preserved by divine will.

Oh, foolish and reckless one, surrounded by disaster! You bring both suffering and the promise of restoration to the world. A leader full of arrogance, a curse upon humanity—who ever wished for you? Who has not suffered because of you?

A mighty king will fall because of you, losing his honorable life. You have turned everything upside down, destroying all that was once beautiful. You have changed the world's order, bringing chaos. Do you try to convince us of your innocence? Do you claim you can justify your actions?

There was once a time when the light of the sun shone brightly upon the prophets, and their words, sweet as honey, were shared with all people. Their wisdom spread like sunlight, bringing knowledge to the world.

But you, the cause of so much suffering, will bring both war and grief. You are the beginning of humanity's struggles and the reason for its suffering. Listen to this terrible prophecy, a warning not to be ignored.

One day, Persia will find peace, free from war and pain. On that day, a divine people, the blessed Hebrews, will pray to God. They will live near the holy city, building strong walls that will rise into the clouds.

The sounds of war will no longer fill their land. No enemy will cut them down, and the righteous will stand victorious over the wicked.

A great man will descend from heaven, a leader above all others. His hands once stretched out upon a sacred tree, the noblest of the Hebrews. He once commanded the sun to stop with his words, speaking with holiness and truth.

So do not fear, chosen people of God. Do not let the sword trouble your hearts. You are his beloved nation, his treasured light, a beautiful and noble branch of faith.

Judea, blessed and full of music, no unholy feet will trample your land again. No more will lawless invaders celebrate on your streets. Instead, your children will honor you with devotion, singing songs of praise in holy tongues.

They will bring offerings and prayers, pleasing to God. Those who endure hardship for righteousness will be rewarded.

The wicked, who spoke against heaven with their evil words, will be silenced. They will hide away, waiting for the world to be transformed.

Flames will rain down from the sky, and people will no longer harvest crops. The land will remain unplanted and untouched until humanity finally recognizes the Lord of all things, the eternal God. They will no longer worship false idols, nor honor the creatures and symbols that Egypt once taught them to revere.

The land of the righteous will be blessed. Sweet water will flow from the rocks, and streams of milk and honey will provide for the faithful. Those who trust in the one true God, the Father of all, will be rewarded for their devotion.

But why has wisdom given me this vision?

Asia, I weep for you and for the lands of the Ionians, Carians, and the gold-rich Lydians.

Sardis, your fate is sealed. Trallis, a city once loved, will fall. Laodicea, a place of beauty, will be shattered by earthquakes, reduced to dust. Darkness will spread across Asia.

The great temple of Artemis in Ephesus will be swallowed by the earth as earthquakes tear through the land. Storms will bring ruin to the sea, sinking ships and leaving Ephesus in mourning. The city will cry out in grief, searching for its lost temple, but it will be gone forever.

Then, the mighty God above will send thunderbolts down upon the wicked. Summer will replace winter in an instant, and disaster will strike. The great Thunderer will wipe out the wicked with fire, lightning, and storms. Their bodies will cover the earth, more numerous than the grains of sand on the shore.

Smyrna will mourn as it approaches the gates of Ephesus, only to meet a worse fate.

Foolish Cyme, ruined by the hands of lawless men, will fall silent forever. She will not even have a voice to cry out. She will remain lifeless, lost beneath the waters of the Cymæan streams.

The people of Cyme will suffer, forced to face the consequences of their actions. They will mourn their city, now reduced to ashes, while Lesbos sinks into the depths of the sea.

Corcyra, a city once full of joy, will be silenced forever. Hierapolis, rich and powerful, will finally receive what it desired—sorrow and endless tears. Tripolis, resting along the waters of the Mæander River, will be swallowed by the waves as God's wrath consumes it in the night.

Miletus, your time will come. A great thunderbolt will strike, bringing destruction from above. Your downfall will come because you misused the wisdom of the god you once followed.

Father of all, look kindly upon the land of Judah, a land full of fruit and prosperity. Let your people witness your justice.

For you, O God, chose this land first, giving it as a gift to humanity, entrusting its people with your divine purpose.

I long to witness the downfall of the Thracians and the destruction of the great wall between the seas, brought low like a river flowing for the fish.

Hellespont, disaster awaits you. One day, a ruler from Assyria will place a yoke across your waters. A great battle will come upon Thrace, stripping it of its strength.

A king from Egypt will take control of Macedonia. But then, invaders from distant lands will rise against the great warriors. The Lydians, Galatians, Pamphylians, and Pisidians will bring destruction, fully armed for battle.

Italy, your time is near. You will be left in ruins, abandoned and unwept, even as your fertile lands remain untouched. A deadly plague will wipe out your people.

And then, high above in the vast heavens, the voice of God will roar like thunder, shaking the world.

The sun's flames will no longer burn, and the bright light of the moon will disappear when God takes control. Darkness will cover the earth, and people will be blinded by fear. Wild beasts and suffering will spread everywhere. This time of sorrow will last long enough for all to understand that God alone rules over everything.

On that day, God will not have mercy on those who worship false gods. People who sacrifice sheep, lambs, goats, and golden-horned bulls to lifeless idols and statues will be judged. Instead of following false beliefs, they should seek wisdom and righteousness. If they fail to love and respect the eternal God, he may destroy all the wicked and shameless people.

At the turning of the moon's cycle, a great war will spread across the world. It will be fought with deception and trickery. From the edges of the earth, a man who killed his own mother will rise to power. He will take control over all lands and rule with unmatched intelligence. He will claim what was once taken from him and bring destruction upon many. He will kill tyrants and burn everything in his path like no one before him. He will even bring back those who were afraid so they can witness his actions.

From the West, war will come, and blood will flow down hills like rushing rivers. In Macedonia, great battles will rage, and a strong force from the West will arrive. But the ruler there will meet his downfall.

Winter winds will blow as war returns to the land, filling the plains with destruction once again. Fire will rain down from the sky along with blood, water, lightning, and thick darkness. A terrible mist of war and death will cover everything, wiping out kings and powerful leaders. When this great battle finally ends, no one will fight with swords, iron, or weapons ever again. These things will no longer be allowed.

Those who survive and learn from past wickedness will finally live in peace, filled with joy.

Murderers of their own families, stop your arrogance and evil ways. In the past, you committed horrible acts—forcing young boys into terrible situations, treating young girls like slaves, and harming the innocent. In your land, even mothers lay with their own children, and daughters were forced to marry their fathers. Kings defiled themselves with wicked acts, and men engaged in unnatural behavior with animals.

You, city of sin and disgrace, will be silenced forever. The time of your celebrations will end. Virgin maidens will no longer tend the sacred fire, and your beloved temple will be reduced to ruins. I have seen with my own eyes the second great house of worship set on fire and destroyed by wicked hands. It was a house that once flourished, a sacred place built by the faithful, believed to be indestructible. But it was cast down.

No one can honor God from the grave. The wise do not worship gold, nor are they deceived by riches. Instead, they honor the one true God with pure offerings and sacrifices. But now, a wicked ruler has risen, leading a great army of powerful men. He destroyed the holy temple and left it in ruins. When he stepped onto sacred land, he defiled it. No one has ever done such a thing before.

Then, from the heavens, a great ruler will descend, blessed by God, holding a scepter in his hand. He will bring justice and return stolen riches to the good and faithful. He will destroy many cities with fire, punishing those who committed evil.

He will rebuild the city that God loves, making it shine brighter than the stars, the sun, and the moon. He will establish order and construct a holy house, pure and magnificent. A mighty tower will be built, stretching up to the clouds, visible to all people.

All those who are righteous and faithful will finally see the glory of the eternal God—a sight they have long waited for. The rising sun will shine upon this new age.

When the final days arrive, people will no longer live in fear. There will be no more betrayals, no forbidden love, no murders, and no chaos—only fairness and justice in all things. It will be the last era of the faithful, when God, the great ruler, completes His plans and establishes a glorious temple.

Babylon, once the greatest kingdom, famous for its golden throne and wealth, will fall. No longer will it stand tall by the Euphrates River. Earthquakes and destruction will bring it to ruins, and the Parthians will cause great suffering. The Chaldeans, known for their strange language, will no longer rule over the Persians and Medes. Babylon, once proud and powerful, will be judged by those it once oppressed.

The very people it ruled will now bring hardship upon it.

In the end, the sea will dry up, and ships will no longer sail to Italy. Asia will be covered in water, Crete will become a flat plain, and Cyprus will face terrible suffering. The people of Paphos will cry over their great loss, and even Salamis, a once-mighty city, will be left in misery. The land will turn to barren sand along the shores, and swarms of locusts will destroy what remains of Cyprus.

When people see the fate of Tyre, they will weep. Phoenicia, your destruction is near. You will crumble into ruins so completely that even the Sirens—those who lure sailors with their songs—will mourn for you.

Five generations after Egypt's downfall, terrible events will unfold. Shameless kings will unite, and the Pamphylians will spread into Egypt, Macedonia, Asia, and Libya. A devastating war will follow, spreading madness across the world. The king of Rome and the rulers of the West will eventually put an end to the bloodshed.

A brutal winter will come, bringing heavy snow and freezing rivers and lakes. A barbaric army will invade Asia, crushing the fierce warriors of Thrace.

In their desperation, people will turn against each other, eating their own family members to survive. Wild animals will invade homes, stealing food from tables. Both beasts and birds will feast on human flesh. The sea will be filled with bodies, its waters turning red with blood. The foolish will suffer, and the earth will grow weak.

A terrible loss will come, leaving so few people that their numbers will be easy to count. They will cry out in sorrow as the sun sets, never to rise again. It will sink beneath the ocean, hiding from the wickedness it has witnessed.

A moonless night will cover the sky, and thick mist will spread across the world once more. But then, a new light from God will shine, guiding those who remained faithful and praised Him.

Isis, once worshipped as a great goddess, will be forgotten. You will be left to wander the waters of the Nile like a lost soul, and no one will remember your name.

Sarapis, once honored with temples of shining stone, will also fall into ruin. Egypt will be left in despair, and those who once followed these false gods will disappear.

Those who mourn for you will cry bitterly, but those who carry wisdom in their hearts and praise God will recognize that you were nothing.

One day, a priest dressed in linen will stand up and say, "Let us build a true and beautiful temple for God. Let us turn away from the old laws of our ancestors, for they did not realize they were worshipping gods made of stone and clay. Let us change our ways and offer praise to the one true God, the everlasting Father, the ruler of all, the giver of life."

Then, in Egypt, a great and pure temple will be built, and the people created by God will bring their offerings there. In return, God will grant them eternal life.

But when the Ethiopians turn away from the sinful Triballians and begin to settle in Egypt, they will bring corruption. In time, they will destroy the mighty temple of Egypt, fulfilling the final prophecy.

God will send his wrath upon the earth, wiping out all the wicked and foolish. No one will be spared because they did not follow what God had given them.

I saw a warning in the sky: the Sun burned fiercely among the stars, and the Moon flashed with an angry light. The stars seemed to struggle against each other as fire battled against the Sun.

Lucifer stood upon the back of Leo, ready to fight. The Moon changed shape, and Capricorn struck Taurus in the neck. Taurus blocked Capricorn from bringing back daylight. Orion broke free from his chains, and Virgo switched places with Aries, taking the fortune meant for Gemini.

The Pleiades stopped shining, and Draco abandoned its usual path. Pisces sank into Leo's belt, and Cancer fled in fear of Orion. Scorpio moved backward into Leo, and Sirius slipped away from the Sun's light.

The great Aquarius blazed with power, while Uranus shook in fury, sending the fighting stars crashing to the earth. When they struck the ocean, flames rose, setting the entire world on fire.

The sky was left empty, with no stars remaining.

Book 6.

I speak from my heart about the great Son of the Immortal, the one praised in songs. His Father gave him a throne to rule over before he was even born. Then, in human form, he was raised up and baptized in the rushing waters of the Jordan River. As he came out of the water, escaping the fire, he was the first to see God's gentle Spirit descending like a white dove with outstretched wings.

A pure flower will bloom, and springs will overflow with water. He will guide people to the right path, showing them the way to heaven and teaching them with wisdom. He will come to bring judgment and try to convince a stubborn people, showing them that he comes from a heavenly Father.

He will walk on water, heal the sick, and raise the dead. He will remove suffering, and from a small amount of food, he will feed many.

When a child is born from the house of David, he will hold power over the entire world—earth, heaven, and sea. He will appear on earth like the first two humans, formed from each other's ribs. This will happen when the world rejoices at the birth of a child.

But for you, land of Sodom, great suffering awaits. You turned against your own God and refused to recognize Him. You crowned Him with thorns, mocking Him, and gave Him bitter gall to drink. Because of this, great troubles will fall upon you.

Oh, blessed wood, the cross on which God was stretched—this earth will not hold you forever. You will rise again, looking toward the house of heaven, when God opens His fiery eyes and reveals His power.

Book 7.

Rhodes, you are unlucky. I grieve for you before any other, because even though you are great, you will be the first to fall. Your people will be lost, but some of your land's riches will remain.

Delos, you will drift on the sea, unstable and uncertain. Cyprus, a mighty wave will one day rise from your shining waters and sweep you away. Sicily, the fire burning beneath you will rise and consume your land.

The people will ignore the warnings sent by God.

Noah alone will survive when disaster strikes. The earth will float, mountains will drift, and even the sky will seem to shift. Everything will be covered in water, and all life will be wiped out. The winds will stop, and a new age will begin.

Phrygia, you will be the first to burn when the waters recede. You will be the first to turn away from God, choosing to follow false gods instead. But your devotion to them will only bring your own destruction.

The people of Ethiopia will suffer greatly. They will be struck down by swords, crying out in pain as they fall to the ground.

Egypt, you who thrive on the riches of the Nile, will turn against yourself. Internal conflict will tear your nation apart. In the chaos, the people will reject Apis, realizing he is not truly a god.

Laodicea, you will never see God, though you act as if you are bold and strong. A great wave from the Lycus River will crash down upon you, bringing ruin.

The mighty God will be born, performing many wonders. He will set an axle in the sky, a terrifying sign for all to see, measuring time with a pillar of fire. The burning drops will fall and destroy those who have committed great evil.

One day, there will be one ruler over all, and people will finally turn to God. But their suffering will not end completely. Everything will come to pass through the house of David, for God Himself has given power into his hands.

Under his command, messengers will rest at his feet. Some will start fires, others will call forth rivers, while some will save entire cities, and others will command the winds.

A heavy burden will weigh upon many, bringing sorrow into their hearts and changing people from within.

When a new branch grows from its roots, creation will once again flourish, providing food for all. The world will be full, but the rulers of that time will be warlike Persians, and their actions will bring terror.

Mothers will take their own sons as husbands, and sons will bring ruin upon their mothers. Daughters will lie with their fathers, breaking the laws of nature.

But then, Rome's warriors will strike with their spears, and blood will flood the land. The leader of Italy will be forced to flee, but his army will leave behind a golden lance, a symbol of their rule, carried by the strongest fighters.

When the ill-fated city of Ilium is finally destroyed, it will not be a place of weddings but of graves. Brides will weep in sorrow, for they did not know God, but instead celebrated with loud music, beating drums and clashing cymbals.

Colophon, seek wisdom from the oracles, for a great fire looms over you.

Thessaly, your land will vanish, and even your ashes will be lost. You will be torn away from the mainland, drifting like wreckage. War will leave you broken, washed away by rushing rivers and swords.

Corinth, your fate is grim. The god of war will surround you, and your people will destroy each other.

Tyre, you will be left abandoned and alone. Once strong, you will be brought to ruin, like a widow grieving the loss of her people.

Cœle-Syria,, last stronghold of the Phoenicians, the sea of Berytus will crash upon you. You are doomed because you did not recognize your God—the one who was baptized in the Jordan River while the Spirit descended like a dove upon him. He existed before the earth and the stars, born from the Father, and took on human flesh through the Holy Spirit. He returned quickly to his Father's house in heaven.

Three great towers were built in heaven, where God's noble guides—Hope, Faith, and Reverence—dwell. They do not find joy in gold or silver but in the righteous acts of people, in sacrifices and pure thoughts.

You will worship the eternal and mighty God, not by burning incense or sacrificing lambs, but by offering prayers with your people. You will release birds into the sky as a sign, sprinkling pure water on the fire, saying:

"As the Father created you, the Word, I send forth this bird as a messenger of my prayers, just as you revealed yourself through fire and water."

You must never turn away a stranger in need. When a hungry traveler arrives at your door, welcome him, sprinkle him with water, and pray three times. Say to God:

"I do not seek wealth. I once received a stranger as a guest, just as I now stand before you, my provider, asking for your mercy."

After praying, give to the traveler, and let him go in peace.

Do not cause suffering, and remain pure in heart and faith. Let the fear of God guide you.

Sardinia, once strong, will turn to ashes. It will no longer be an island when the tenth age comes. Sailors will search for it, but it will be gone, and seabirds will cry over its disappearance.

Mygdonia, trapped by the sea, you will exist for many ages, but in the end, a scorching wind will destroy you, bringing countless sorrows.

Celtic lands, high in the mountains beyond the Alps, deep sand will bury you completely. You will no longer provide grain or livestock, and ice will cover your land, punishing you for the wrongs you failed to recognize.

Mighty Rome, you will send out lightning-like power after the Macedonian spears, but God will make you disappear when you believe yourself to be strongest. You will suffer and cry out in agony.

Rome, I warn you once more—your downfall is certain, and you will not stand forever.

Syria, I grieve for you. Your future is filled with suffering, and I mourn what is to come.

O people of Thebes, you do not see the danger ahead. A terrible sound is coming, while flutes play their tunes.

A trumpet will blast a warning, and you will witness your land being destroyed.

Woe to you, suffering city! Woe to the raging sea! Fire will consume you completely, and the saltwater will bring destruction to your people.

A great fire will spread across the earth like a flood, burning everything in its path. The mountains will be set ablaze, rivers will dry up, and springs will vanish.

The world will fall into chaos as people perish. Those who are burned will look up, but instead of stars, they will see only fire in the sky.

Their suffering will not end quickly. Their bodies will waste away, and their spirits will burn for many years. Then, they will understand that God's law is not to be tested or ignored.

The earth will be filled with suffering because people worshipped false gods and filled their altars with lies.

Those who tell false prophecies for money will face great pain. The Hebrews, dressed in rough sheep's wool, will also prove false. They will not receive what they expected, but instead, they will change their ways and no longer mislead the righteous—those who truly worship God with their hearts.

After many years, in a new era, the world will change.

Darkness will cover everything, and the night will be long and without light. A horrible stench of burning sulfur will fill the air, signaling the deaths of many, who will perish by violence and starvation.

Then, God will create a pure mind in people and restore the world to how it was meant to be.

No one will need to plow fields, and no oxen will pull the plow. There will be no need for crops or vineyards, because all people will eat together from the food God provides, like sweet manna from heaven.

And God will live among them, teaching them, just as he has taught me.

I have done many terrible things, some knowingly and some in ignorance.

I lived a life of selfishness, ignoring the laws of marriage and breaking sacred oaths.

I turned away those in need and followed my own desires, ignoring God's words.

Because of this, fire has consumed me, and I will suffer.

I will not live forever, but in time, my punishment will come. By the sea, people will build a tomb for me, and I will be stoned to death.

For I have committed a terrible sin—bringing forth a son by my own father.

Strike me down! End my life! Only then will I be able to lift my eyes toward heaven.

Book 8.

God's warnings of great anger and destruction in the final days will come upon an unfaithful world. I reveal these things to all people, speaking to every city.

From the time when the great tower fell and human speech was divided into many languages, different kingdoms rose to power. First was Egypt, then Persia, Media, Ethiopia, Assyria, and Babylon. After them came the proud rule of Macedonia. Finally, the fifth and last kingdom, that of the Italians, will bring great suffering upon the earth. It will force people from all nations into its control, create laws for many, and claim power over all things.

But even the strongest rule will not last forever. God's justice may take time, but in the end, everything will be reduced to dust. Fire will destroy all things, burning even the highest mountains and consuming every living being.

The root of all evil is greed and foolishness. People will love gold and silver more than anything else. They will treasure wealth above the light of the sun, the sky, the sea, or the land that provides them with food. They will forget God, the true giver of all things, the Father of all. Faith and kindness will be ignored.

Greed will bring chaos, turning families against each other. With money as their guide, people will no longer honor marriage. Lands and seas will be divided, guarded by those who serve the wealthy. The powerful will take from the hardworking, pretending it is fair, while expanding their own wealth.

If the earth were not placed far from the heavens, even the light would be bought and sold. The rich would claim the sun itself, and God would have to create another world for the poor.

But Rome, your time will come. A punishment from above will strike you. You will be the first to fall, brought low and turned to ruins. Fire will consume you, reducing your streets to ashes. Your treasures will be lost, and wild animals will take over your empty foundations. You will be completely abandoned, as if you had never existed.

Where will your sacred statues be then? What god will save you—gold, stone, or bronze idols? Where will your rulers and laws go? Where will the proud bloodline of your gods—Rhea, Cronus, and Zeus—be? You worshipped lifeless images of the dead, giving honor to forgotten graves, while Crete foolishly took pride in these empty tombs.

Rome will have fifteen rulers who spread their power from the east to the west, enslaving nations. Then, an old man, whose name is tied to the sea, will rise. He will travel the world, gathering wealth, stealing gold and silver wherever he can.

He will take part in false religious rituals, declare his child to be a god, and erase all sacred things. He will expose old lies for what they are. But his own time will come, and his death will bring sorrow to the land.

People will see that destruction is near. They will realize that Rome's strength is crumbling. Parents and children will cry out together, mourning along the banks of the Tiber River, knowing the end is near.

After him, in the final days, three rulers will rise. Their names will reflect the power of God, who rules now and forever. One of them, an old man, will hold the throne for a long time. He will store up the world's wealth, preparing for the return of a fugitive who once killed his own mother. When that man comes back from the farthest parts of the earth, the old ruler will distribute his riches and make Asia prosperous.

Then, Rome, you will mourn. You will take off your royal robes and wear clothes of sorrow. Your pride will be gone, and you will never rise again. The mighty legions that once carried the eagle into battle will fall.

Where will your power be then? What lands will still follow your reckless rule? The whole world will be in chaos when the Almighty comes to judge the living and the dead.

Families will turn against each other. Parents and children will no longer love one another because of their faithlessness and suffering. You will face grinding despair, your cities will fall, and the earth will shake and split open.

A fiery dragon with glowing eyes and a full belly will rise from the sea, bringing destruction. Your people will suffer famine and civil war. When that happens, the end of the world will be near. The final day of judgment will come, and God will choose those who are worthy.

The wrath of God will strike Rome first. There will be endless bloodshed, and life will become unbearable.

Oh, foolish and reckless nation, you never understood where you came from or where you are going. You entered this world naked and unworthy, and you will return to nothingness before facing judgment for your unjust rule.

A mighty force will come down from the heavens, and you will be cast deep into the earth. Flames, burning oil, and sulfur will consume you. You will vanish into fire and become dust for eternity. Those who see it will hear the cries of suffering rising from the depths of the underworld—shouts of grief and teeth grinding in pain.

In death, there is no difference between rich and poor. Everyone leaves the earth as empty-handed as they arrived. There will be no kings, rulers, or tyrants. No judges will take bribes. No blood will be poured on altars. There will be no drums, no flutes, no wild dancing, no music of harps or trumpets calling to war. The dead do not fight or argue. They do not carry swords.

The afterlife is a prison, locked for eternity, waiting for God's final judgment.

Rome, your golden idols will not save you. You will face your first punishment, and your people will weep and grind their teeth in agony. No longer will Greeks, Syrians, or any foreign nation bow to your rule. Instead, you will be plundered and forced to suffer the same oppression you once inflicted on others. You will be left in ruins, a shameful reminder to the world.

Then, the sixth generation of Latin rulers will come to an end, and their line will disappear. A new king from the same land will rise, ruling over every nation. He will have full power, and his descendants will continue to rule, as destined by God.

Time will continue moving forward, and when Egypt has been ruled by fifteen kings, a great event will take place.

The era of the Phoenix will come to an end, and a fierce, chaotic army will rise against the Hebrews. War will rage, and Rome's pride will finally be shattered.

Rome's power will fall even while it still seems strong. Once a mighty queen among cities, it will no longer thrive when a ruler from Asia arrives, bringing war. After he has completed his conquests, he will come to the city.

When Rome has ruled for 348 years, disaster will strike. The city will be taken by force, and its name will be erased. I, so full of sorrow, wonder if I will live to see the day of Rome's destruction—a day that will bring the greatest suffering to the Latin people.

A leader with a fiery spirit will come from Asia, riding in a Trojan chariot and hiding his children. When he cuts through the land and crosses the sea, destruction will follow. Blood will spill as the mighty empire falls. A dog will chase a lion, and the shepherds will be destroyed. His power will be taken from him, and he will descend into the underworld.

Rhodes will suffer its worst disaster. Thebes will fall under cruel rule. Egypt will collapse under corrupt leaders.

A man who somehow escapes this destruction will be truly blessed. Rome will be reduced to ruins. Delos will lose its beauty, and Samos will turn to sand.

Pride will bring disaster upon the Persians, and their arrogance will be crushed.

Then, the holy ruler of the world will raise the dead and take control forever. Rome will face three terrible punishments from God. People will be destroyed by their own actions because they refused to listen and change their ways.

When famine, plague, and violent war increase, the former ruler will call the senate and plan even greater destruction.

The earth will bloom again, and rainstorms, fire, and strong winds will sweep over the land. But people will continue in their shameless ways, ignoring the warnings of both God and men. They will follow greedy and corrupt rulers, seeking wealth without end. They will lie, betray, and destroy faith, showing no sense of right or wrong.

In time, the stars will fall into the sea, one by one. A brilliant comet will appear in the sky, a warning of the war and suffering to come.

I do not want to live when a wicked woman rules. But the day will come when divine grace reigns, and a holy child will defeat the great destroyer of mankind. The depths of the earth will be opened, and suddenly, the world will be covered.

When the tenth generation has passed into the underworld, a woman will rise to power. Her rule will bring hardship, and God will allow even greater suffering.

The sun will grow dim, shining weakly even at night. The stars will disappear, and great storms will shake the earth.

Then, the dead will rise. The lame will run, the deaf will hear, the blind will see, and the mute will speak. Life and wealth will be shared among all people.

The land will not be divided by walls or fences, and it will produce more than ever before. Streams of sweet wine, milk, and honey will flow freely.

Then, the judgment of God will come.

Seasons will change—the winter will bring summer. When God decides, all prophecies will be fulfilled, and the world will come to an end.

When the time of judgment comes, the earth will tremble and sweat. From heaven, the eternal King will appear to judge all people and the entire world.

Both the faithful and the unfaithful will see God standing with the saints at the end of time. He will judge every soul, and the world will become empty and covered in thorns. People will throw away their idols and wealth, realizing they are worthless.

A great fire will consume the earth, the sky, and the sea, even breaking open the gates of the underworld. The dead will rise and stand in the light alongside the saints, but the wicked will be trapped in the fire, suffering for eternity.

Every secret will be revealed, as God will expose the hidden thoughts of all people. Many will cry out in sorrow, grinding their teeth in regret. The sun will darken, the stars will stop shining, and the sky will be rolled up like a scroll. The moon's light will fade, mountains will be flattened, and hills will disappear. The earth will become level, and the seas will vanish. Rivers and springs will dry up as the world is scorched by intense heat.

A trumpet will sound from heaven, filling the world with a terrifying noise, warning of the suffering to come. The ground will split open, revealing the depths of the underworld, and all kings will stand before God's judgment seat. Fire and burning sulfur will pour down from the heavens.

A great sign will appear—a holy symbol recognized by believers, but rejected by the world. It will bring light to those who follow the truth and mark the faithful with its power. Twelve springs of water will flow, and a strong shepherd will guide the people with an iron staff.

The hidden message written in acrostics reveals the name of the Saviour, the immortal King who suffered for the sake of the world.

Moses foreshadowed him when he lifted his hands in faith, helping his people win the battle against Amalek. This showed that he was chosen and honored by his Father, God.

He is the rod of David, the promised foundation. Those who believe in him will have eternal life.

He will not come in glory but as a humble man, without honor or beauty, to bring hope to those who suffer. He will restore the beauty of human life and bring faith to those who have none. He will give new life to the first man, who was created by God's hands but was led astray by the serpent's deception, bringing death and the knowledge of good and evil. Turning away from God, humanity followed its own ways instead.

From the beginning, the Almighty spoke to his Son, saying, "Let us create mankind in our image." God shaped them with his hands, and in time, the Son would bring them back to their true form. Keeping this promise, he will enter the world, born of a holy virgin, and he will be baptized with water by the hands of elders. Through his words, he will perform miracles and heal every sickness.

With just his voice, he will calm the wind, and with his footsteps, he will bring peace to the raging sea. From five loaves and a fish, he will feed five thousand people in the desert, gathering twelve baskets of leftovers as a sign of hope for all nations.

He will call upon the souls of the faithful and show love to those who suffer. Though he will be mocked, beaten, and rejected, he will repay evil with kindness, embracing a life of humility. He sees and hears all, searching the hearts of people and revealing the truth. He is the Word that created all things, and through him, even the dead will rise, and every sickness will be healed.

But in the end, he will be handed over to cruel and faithless men. They will strike him with their hands and spit on him with mouths filled with hatred. He will be whipped and remain silent, revealing nothing about himself, even as they mock him.

They will place a crown of thorns upon his head, for the thorns are a symbol of his eternal kingship. They will pierce his side with a reed, fulfilling the law, though their hearts are ruled by anger and revenge.

When all these things happen, every law written by men will be fulfilled. He will stretch out his hands, embracing the whole world. But instead of kindness, they will give him gall for food and vinegar to drink.

The curtain in the temple will tear in two, and in the middle of the day, darkness will cover the earth for three hours.

The old way of worship, hidden behind rituals and traditions, will no longer be needed when the Eternal One walks among people. He will enter the depths of the underworld, bringing hope to the faithful and announcing the end of time. He will fall asleep in death, but on the third day, he will rise again, breaking the power of death.

Emerging from the grave, he will be the first to show the way to resurrection. He will cleanse people's sins with the waters of life, so they may be reborn and no longer enslaved to the ways of the world.

First, he will appear to his followers in the flesh, just as before. The marks of his suffering will remain on his hands and feet, representing the four corners of the world. The leaders of the earth will have carried out a terrible and unjust act against him.

Rejoice, daughter of Zion, for though you have suffered, your King is coming! He will arrive humbly, riding on a young donkey. He will break the heavy chains of oppression and set people free from unjust laws and burdens.

Know your God, the Son of God. Honor him, keep him in your heart, and love him with all your soul. Praise his name and turn away from your old ways. Wash yourself clean of the sins of the past. He is not pleased by songs or prayers alone, nor does he care for sacrifices of things that perish. Instead, offer him a sincere heart and true understanding. When you do, you will see the Father.

Then, the world will be left in silence. The air, the land, the sea, and the light of the sky will stand still. Day and night will blend together into emptiness. The stars will fall from the sky. Birds will no longer fly, and animals will disappear. No voices of people, birds, or creatures will be heard. The world will be without sound, except for the deep sea roaring in anger. The creatures of the ocean will die, and no ships will sail the waves.

The earth will be soaked in blood from endless wars. People will cry out in pain, grinding their teeth in fear and suffering. Hunger, thirst, disease, and violence will consume them. They will wish for death, but it will not come. There will be no rest, no night to bring relief. They will cry out to God for help, but he will turn away, for he gave them many chances to repent.

God has shown me all of this, and everything I have spoken will come to pass. I know the number of grains of sand, the depths of the sea, the hidden places of the earth, and the darkness of the underworld. I know the stars, the trees, every living creature, and every person—those who have lived, those who live now, and those yet to be born. I gave people the ability to think, see, and hear. I understand even the silent thoughts of the heart.

I know everything, from the beginning to the end. I alone am God, and there is no other. But people carve statues from wood and call them gods. They shape them with their hands and bow before them, singing praises to lifeless idols and performing meaningless rituals.

People turned away from their Creator and became slaves to sin. Even though they had everything they needed, they wasted their gifts on things that could not help them. They offered sacrifices to lifeless idols, burning flesh and bones as if it were for their own dead. They poured out blood on altars for demons and lit fires in my name, thinking I needed light. They foolishly believed I was thirsty, offering wine to statues that could not drink or save them.

I do not need your burnt offerings, your incense, or your sacrifices of blood. These rituals were created in honor of kings and tyrants, not for me. They worship false gods, calling lifeless statues divine while forgetting the Creator who gave them life. They put their trust in things that cannot see or hear, completely blind to what is truly good.

I gave people two paths—one leading to life and the other to death—and I warned them to choose the path of goodness. But they ran toward destruction, bringing eternal fire upon themselves.

People were made in my image, with reason and wisdom. Instead of offering bloody sacrifices, prepare a pure table. Feed the hungry, give water to the thirsty, clothe those in need, and help those who suffer. Support the weak and provide kindness, for that is the true sacrifice I desire. If you do this, I will reward you with eternal light and a life that never fades when I cleanse the world with fire.

I will test everything, separating what is pure from what is corrupt. I will roll up the heavens, open the depths of the earth, and raise the dead. I will end fate and destroy the power of death. Then I will bring judgment, testing both the righteous and the wicked. I will separate the faithful from the unfaithful, just as a shepherd separates sheep from goats. Those who sought power and silenced the righteous will be cast out.

No longer will anyone worry about tomorrow or dwell on the past. Time will not be measured by days, seasons, or sunsets. Instead, I will create a never-ending day filled with the light that my people have longed for.

You, the eternal and pure one, measure the power of storms and command the lightning. You calm the roar of thunder and control the raging winds. You hold back the fire of destruction and soften the blows of harsh weather. You know everything that will happen and decide what is good.

Your Son, who has always been with you, agrees with your will. Together, you created life and gave humans breath. You said, "Let us make man in our image and give him dominion over the world." And so, by your word, everything was made—the heavens, the air, the fire, the earth, the sea, the sun, the moon, and the stars.

Both day and night, in sleep and while awake, in spirit and thought, in strength and wisdom, all living creatures—those that swim, fly, walk, or crawl—were created by his will, following your guidance.

In the last days, he passed through the earth, coming from the womb of the virgin Mary. A new light appeared, and from heaven, he took on human form. First, the angel Gabriel came in his shining presence and spoke to the young woman, saying,

"O pure virgin, receive God into your heart."

As he spoke, he brought God's grace upon her. She was filled with fear and amazement at his words, trembling with surprise. Her heart raced, and she was overwhelmed. But soon, she felt comforted, her heart was lifted, and she smiled with joy. A soft blush appeared on her face, and she felt a sense of wonder and humility.

Then, the Word entered her womb, and in time, he took on flesh and life, becoming a human child through a miraculous birth. This was a great wonder for the world, but not for God the Father or God the Son.

The earth rejoiced at his birth, the heavens celebrated, and the world was filled with joy. A new star appeared in the sky as a sign, and wise men honored it. The newborn child was revealed in a humble manger to those who followed God—shepherds, goatherds, and those who cared for the flocks. Bethlehem, chosen by God, became the birthplace of the Word.

Live with humility in your heart. Reject cruelty and love your neighbor as yourself. Serve God with all your soul.

We, who come from the holy line of Christ, are called as one family, united in faith. We live in truth and righteousness. We do not enter the temples of false gods, offer sacrifices to idols, or worship lifeless images with prayers, incense, or candles. We do not bring offerings of bulls or shed the blood of animals as if such things could erase sin. We do not burn the flesh of sacrifices, sending foul smoke into the sky.

Instead, we worship with pure hearts, filled with love and kindness. With joyful songs and praise, we honor you, the eternal and true Father of all, full of wisdom and light.

Book 11.

O people of the world, scattered far and wide, living in great cities and powerful nations, spread across the east, west, north, and south, speaking many different languages and ruled by many kings—I must tell you of terrible things to come.

A long time ago, after the great flood wiped out the first people, God created a new race of humans. But they became proud and tried to build a tower so high that it would reach heaven. In his anger, God confused their languages and brought chaos upon them, causing them to fight among themselves. Since then, ten generations of people have lived on earth, spreading across different lands and forming separate nations.

First, Egypt will rise to power, ruling with justice. Wise men will govern the land, but eventually, a fierce and ruthless leader will take control. His name will be marked by the first letter of an acrostic, and he will use his sword against those who remain faithful to God.

During his rule, a great sign will appear in Egypt—one that will bring hope. Crops will grow in abundance, saving people from terrible famine. A great lawgiver, once a prisoner, will come from the east, born from the people of Assyria. His name will carry the number ten.

When Egypt is struck by ten great plagues sent from heaven, I will speak again of these things.

Memphis, a great city, will suffer destruction. The waters of the Red Sea will swallow many of its people.

Then, when the twelve tribes of Israel leave the land of Egypt as God commands, the Almighty will give them laws to guide all of humanity. A mighty king will rise to rule over them, a strong and noble leader whose name will be linked to Egypt and Thebes. Though he will pretend to be kind, he will be a dangerous ruler, watching over his people while preparing for war.

After twelve generations, lasting over seven hundred years, Persia will take control. Then, suffering will fall upon the Jewish people—a time of terrible famine and disease that they will not escape.

A Persian king will rule, and after him, his grandson will inherit the throne. But when 549 years have passed, Persia will collapse, and its people will become slaves to the Medes. They will suffer greatly in battles and be struck down by their enemies.

Disaster will come quickly to the Persians, Assyrians, Egyptians, Libyans, Ethiopians, Carians, Pamphylians, and many others.

A ruler will pass his power to his grandsons, who will take control of the entire world, stealing from many nations without mercy. The Persians will mourn by the Tigris River, and Egypt will weep as its people suffer.

Then, a wealthy man from India will bring destruction to the land of the Medes. The people of Media will pay for their past sins and will be forced to serve Ethiopian rulers for 107 years, carrying heavy burdens under their rule.

After this, an Indian king with dark skin and gray hair will take power. He will be a strong and determined leader, bringing terrible wars to the East. His rule will be harsh, and he will wipe out many people. For 30 years, he will hold his throne, and after 17 more, nations will rise against him, seeking freedom. For three years, they will fight for their independence. But he will return, forcing all the nations back under his control, making them serve him once again. Afterward, peace will spread across the world.

Then, a great Assyrian king will rise, a leader who will persuade people to follow the laws of God. The most powerful rulers will fear him and submit to his wisdom because he will rule with understanding and fairness. He will rebuild the temple of God and destroy idols, bringing people together, including the elderly and young children. His name will carry the number 200 and a sign of the 18th letter.

For 25 years, he will lead, but when his time ends, many kings will rise—one for each tribe, city, island, and land. Among them, one will be greater than the rest, ruling over the other kings and their descendants for 80 years before their time ends.

When a fierce and violent ruler appears, bringing war, destruction will return to the Persian land. The rivers will run red with blood when a stronger conqueror arrives.

Italy will bring forth a new ruler, a powerful force that will shock the world. There will be the cries of young children near a pure spring in a dark cave, born from a wild beast that feeds on sheep. As they grow, they will build a mighty city upon seven hills and lead wars, destroying many people. Their names will become a great sign for future generations.

They will construct great walls around their city and fight many wars. In Egypt, there will be rebellion and suffering, as I have warned before. A terrible disaster will strike Egyptian homes, and once again, its people will turn against each other.

I grieve for you, Phrygia. A conqueror from Greece, known for taming horses, will invade your land, bringing war and disease.

Troy, I pity you. From Sparta, a powerful and vengeful woman will come to your city, bringing ruin. She will bring endless suffering, filling your streets with pain and sorrow. When the Greeks attack, the bravest warriors will rise to battle, and among them, a strong king will fight for his brother, carrying out terrible acts.

The mighty walls of Troy will fall. After ten long years of bloody war, a clever trick will bring its downfall. A great wooden structure will be taken into the city, but the Trojans will not realize it hides Greek warriors inside. That night, disaster will strike. So many lives will be lost in a single night, and an old man will weep over the destruction. But even in ruin, Troy's name will live on in history.

A great hero, born from Zeus, will have a name beginning with the first letter of the alphabet. He will return home in triumph, only to be betrayed and killed by a deceitful woman.

Then, a new leader will rise from the bloodline of Assaracus, a strong and brave man. He will escape the fires of Troy, carrying his elderly father on his shoulders and leading his son by the hand. As he flees, he will carefully avoid the raging flames and pass through the dangers of both land and sea. His name will have three syllables, and the first letter of the alphabet will mark him as an important figure.

He will establish a city for the powerful Latin people, but in his fifteenth year, he will meet his end, drowning in the sea. Even in death, he will not be forgotten, and his descendants will go on to rule vast lands stretching to the Euphrates and Tigris rivers, across the heart of the Assyrian kingdom and the lands of the Parthians.

A wise old man, a great poet, will also arise. He will be known for his knowledge and wisdom, writing incredible stories that will shape the world. Some of his words will even come from my own verses. He will reveal my books but then hide them away, keeping them secret until the final days of life and death.

When these events unfold, the Greeks will once again fight among themselves. The Assyrians, Arabs, Medes, Persians, Sicilians, Lydians, Thracians, and Egyptians will all be drawn into war. Chaos will spread across the lands, and God will bring confusion to all nations.

A ruthless leader from Assyria, with the heart of a beast, will rise to power. Cunning and brutal, he will cross every land and sea, striking down all in his path. Greece, once strong, will suffer greatly.

For 78 years, Greece will endure war and suffering, reduced to ruins as it becomes the battlefield for many nations.

Then, a Macedonian ruler will rise, bringing even more pain to Greece. He will destroy Thrace, the islands, and the warlike Triballi. A powerful warrior, he will bear a name linked to the number 500. His rule will be short, but he will leave behind the greatest empire on earth. However, he will meet his end by a soldier's spear while believing he was safe.

His son, a strong-willed child, will take the throne. His name will start with the first letter of the alphabet, but his dynasty will not last. Though some will claim he is the son of Zeus or Ammon, he will not be seen as their true heir.

This ruler will lead many wars, seizing cities and bringing suffering to countless people. His conquests will leave a deep wound on Europe. He will devastate Babylon and every land under the sun, becoming the only man to conquer both the East and the West.

Oh, Babylon, your time of glory is over. Once called a queen, you will now serve the victories of others. War will come upon Asia, bringing great destruction. Your people will suffer, and many will be slain.

A powerful warrior, known by the number four, will rise. He will be skilled with the spear, fearsome in battle, and deadly with the bow and arrow. But soon after, famine and war will spread across Cilicia and Assyria. Kings will fight each other, consumed by endless conflict.

You must flee from your former ruler—do not stay, but do not be afraid to leave. A terrible beast, like a raging lion, will appear. He will be a brutal leader, cruel and without justice, wearing a cloak upon his shoulders. Stay away from the man who wields thunder as his weapon. All of Asia will fall under harsh rule, and the land will be soaked in blood.

A powerful leader from Pella will establish a mighty city in Egypt, named after himself. But despite his greatness, he will meet his end, betrayed by his own companions. This leader will be murdered at a feast after leaving India and returning to Babylon.

After him, other rulers will rise—cruel, selfish men who will devour their people. They will each rule over their own tribes, but none will bring peace. Then, a mighty leader will come, uniting all of Europe. However, after much bloodshed, he will surrender to fate and die.

Following him, eight rulers will come from the same family, all sharing the same name.

At this time, a queen from Egypt will rise to power. She will rule a great city, Alexandria, the pride of the Macedonian empire. It will shine brightly as the center of civilization. But Memphis will resent her rule. There will be peace in the world, and Egypt will become more fertile than ever, bringing forth abundant crops.

However, disaster will strike the Jewish people. They will face famine and an unbearable plague with no escape. But Egypt, with its rich land, will provide refuge for many wandering souls.

Egypt will have eight kings, ruling for 233 years. But their family line will not last. A woman will rise, bringing ruin to the kingdom, betraying her people. One by one, the rulers will fall—fathers killed by sons, and sons murdered before they can bear their own children. Yet, a new ruler will eventually emerge, and a new dynasty will grow.

A queen will take the throne, ruling over the Nile, which flows into the sea through seven streams. Her name will have the number twenty. She will demand great riches, gathering gold and silver. But her own people will betray her. Egypt will once again be thrown into war, filled with bloodshed and destruction.

Meanwhile, Rome will be ruled by many leaders, none of them truly great. Tyrants will take power, with thousands of rulers and countless officials controlling the laws and assemblies. The mightiest emperors, known as Caesars, will govern, but they will all meet tragic fates.

The last of these emperors will have a name marked by the number ten. He will fall to war, struck down by an enemy. The young men of Rome will carry his body and bury him with honor, offering tributes in his memory.

When 620 years have passed since Rome's founding by its legendary ruler, no longer will dictators rule for fixed terms. Instead, a single ruler will take full control, a king who will be seen as equal to the gods.

Then, Egypt, prepare for the king who will come to you. A fierce warrior with a shining helmet will arrive, bringing war.

You, once powerful, will be conquered and left defenseless. Battles will rage around your cities, bringing destruction. After suffering greatly in these wars, you will be forced to submit, unable to resist. In the end, you will join with a powerful ruler, bound together in an unwanted union.

Oh, unfortunate bride, you will surrender your kingdom to the Roman ruler. You will pay for everything you once did when you ruled with strength. Your entire land, stretching to Libya and the lands of dark-skinned people, will be given to this unstoppable leader as part of your dowry.

No longer will you stand alone, for you will be joined to a ruthless warrior—a brutal, merciless conqueror. But this will not bring you happiness. You will be forgotten, your name lost among the people. Your once-proud legacy will vanish, and a tomb will encircle you, a resting place for a ruler who has fallen.

A great crowd will mourn you, and even the mighty king will grieve your passing.

Then, Egypt, you will become a servant, forced to bear the burden of war against the Indian lands. You will be ruled harshly, and the Nile will flow with your sorrow like endless tears. Though once a land of riches, feeding great cities, you will now provide for cruel invaders, men who show no mercy.

Oh, wealthy Egypt, how many invaders will claim you as their prize? Once home to powerful rulers, you will now be a slave to foreign nations. You will suffer for what you once did to others—when you enslaved a people who worshiped God, forcing them to toil under the sun, breaking their backs for your wealth. You made them cry tears that watered your fields.

Because of this, the eternal God, who rules from heaven, will bring judgment upon you. You will pay for the wrongs you committed long ago, and only then will you understand that God's wrath has fallen upon you.

Now, I will go to Python and Panopeus, cities with great towers, where people will finally recognize that my words are true. They will no longer doubt my prophecies.

When you read these words, do not be afraid. Everything that is to come, and everything that has already happened, will be revealed. Then, no one will say that this prophecy was given by chance.

Lord, let this be the end of my song. Take away the madness that fills my voice and replace it with a song of peace.

Book 12.

Listen now to the sorrowful days that will come for the people of Latium.

First, after the kings of Egypt have fallen and been buried in the earth, and after the man from Pella, who ruled over both the East and West, has died—his body abandoned by Babylon, proving false the claims that he was the son of Zeus—there will come a ruler from the bloodline of Assaracus, a descendant of Troy. He will survive great destruction and fire.

Many rulers will follow—some great warriors, others young and inexperienced. After six hundred and twenty years of Rome's rule, the first great leader from across the western sea will take control. He will be a powerful and warlike ruler whose name begins with the first letter of the alphabet. He will conquer lands rich in food and bring devastation through war. The lands he conquers will pay for the wrongs they have done.

This great warrior will be the strongest in battle. Thrace and Sicily will submit to him, and Memphis will fall, its downfall caused by corrupt leaders and a powerful, independent woman who will die by the sword. He will establish laws for the people and bring everything under his control. His rule will be long and filled with great fame, and no king before or after him will be greater.

During his time, the world will witness signs—marvelous seasons and wonders on the earth. And when a bright star, shining like the sun, appears in the sky at midday, the hidden Word of the Most High will come down, taking human form. The power of Rome and its people will grow stronger. But in time, this mighty king will die, passing his rule to another.

The next ruler will be a strong warrior, wearing a purple cloak. His name will carry the number three hundred, and he will lead great battles. He will defeat the Medes and the Parthians, who fight with arrows. He will bring destruction to the great city, and Egypt, Assyria, and distant lands near the Rhine will suffer under his rule. He will also attack a city near the river Eridanus, known for its evil schemes. But in the end, he will be struck down by a shining blade.

After him will come a cunning ruler, whose name carries the number three. He will amass great wealth, but his greed will have no limits. He will continue taking more and more from the earth, never satisfied. However, there will be a time of peace, and wars will cease. He will seek wisdom through divination, hoping to secure his rule. But a great sign will appear—while he is dying, small drops of blood will fall from the sky.

He will bring great suffering to the people of Rome, committing terrible crimes. He will kill the leaders of the assembly and cause a severe shortage of food. The people of Cappadocia, Thrace, Macedonia, and Italy will struggle through a devastating famine.

Egypt will be the only land able to feed many people. Meanwhile, a deceitful king will secretly harm an innocent young woman. The people, filled with sorrow, will give her a proper burial and turn against the king, plotting against him in anger. While Rome is still strong, this ruler will fall.

A new leader will take power, marked by the number twenty. He will bring war and suffering to the Sauromatians, Thracians, and the Triballi, fierce warriors known for their skill with spears. Rome will spread destruction, and a terrifying sign will appear when this ruler governs Italy and Pannonia. At midday, darkness will cover the sky, and stones will fall from heaven. Soon after, the ruler of Italy will meet his fate and die.

Then, a new leader, known by the number fifty, will rise. He will bring destruction to the wealthy and powerful, acting like a venomous serpent spreading war. He will even bring harm to his own family and cause chaos everywhere. As a chariot racer, he will take part in deadly contests, spilling blood and leaving destruction behind. He will cut through the land between two seas, staining it with blood. But in the end, he will disappear, only to return, claiming to be a god. However, the true God will prove him powerless.

During his reign, peace will come for a time, and people will live without fear. Water will flow through the land, creating new paths. The ruler will organize grand games and contests, even competing himself, singing and playing music. Eventually, he will abandon his throne and die far from home, facing the consequences of his actions.

After him, three rulers will rise. Two of them will have names linked to the number seventy, and another will be connected to the third letter of the alphabet. They will meet violent ends in different places, falling in battle. Then, a mighty warrior, known by the number seventy, will take power. He will bring suffering to the faithful, leading brutal campaigns against Phoenicia and Assyria. His armies will bring war to the sacred land of Solyma and reach the shores of the Tiber.

Phoenicia will suffer greatly, bound by its own victories, and other nations will trample over it. Assyria will fall, and its people, including children and wives, will be taken as captives by foreign rulers. Their wealth and way of life will be destroyed. God's wrath will come upon them for abandoning His laws and worshipping false gods. Wars, famine, and disease will spread across the land.

Eventually, this merciless king will face his fate. After him, two rulers will take power, honoring their great father. They will become warriors, leading armies and continuing the fight. One of them will be a noble leader, whose name equals three hundred, but he will be betrayed and killed—not in battle, but by treachery in Rome.

Then, a ruler marked by the number four will take control. He will be beloved across the world, bringing peace. Nations from the west to the east will willingly follow him, and cities will submit to his rule without force. The world will rest from war, and he will be blessed by the Almighty God.

However, famine will strike Pannonia and the lands of the Celts, taking many lives. In Assyria, where the Orontes River flows, great buildings will rise, admired by the powerful ruler. But his trust in others will lead to his downfall—he will be wounded in his own palace, betrayed by a friend, and die unexpectedly.

After him, another ruler will rise, linked to the number fifty. He will bring destruction to Rome, taking the lives of many citizens, though his reign will be short. His fate will be sealed because of a past ruler's actions.

Then, a new leader, known by the number three hundred, will take the throne. He will launch fierce attacks on Thrace, Germany, and Iberia, spreading destruction. The Jews will suffer another great tragedy, and Phoenicia will be soaked in blood. The walls of Assyria will crumble as invaders tear through the land, leaving behind ruin and death. Finally, another ruthless ruler will rise, bringing even more devastation.

Then, disasters from the mighty God will come—earthquakes, plagues, unexpected snowstorms, and powerful lightning strikes will affect every land.

A great Celtic king, who roams the mountains, will meet an unfortunate end as he eagerly rushes into battle. Worn out from war, he will fall, and foreign soil will cover his body—soil named after Nemea's flower.

After him, another ruler will rise, an older man with silver hair. His name will have four syllables and be linked to the sea, with the letter A from Ares at the start. He will build temples in many cities and travel the world, bringing back riches like gold and amber. He will rid the sacred places of magicians and replace them with better things for people. While he rules, there will be great peace, and he will be known for his strong voice, fairness, and commitment to justice. But in the end, he will bring about his own downfall.

Following him, three rulers will take the throne, with the third one ruling for three decades. After them, a king from the first unit of numbers will take power, followed by another leader whose number will be seven tens. Their names will be honored, and they will defeat many people, including the Britons, Moors, Dacians, and Arabians.

When the last of these rulers falls, war will break out again. Ares, who was once wounded, will return to battle and destroy the Parthians completely. However, the king himself will be betrayed and killed by a wild beast that was trained by his own hands.

After him, a new ruler will take the throne. He will be wise and skilled in many ways, and his name will be the same as the first great king from the first unit of numbers. He will be strong and do great things for the Latins in memory of his father. He will decorate Rome with gold, silver, and ivory, and walk proudly through the markets and temples with another strong leader.

But soon, terrible wounds of war will spread across Rome. He will launch an attack on the German lands, and during this time, a great sign from God will appear in the sky. God, who sees all, will send unexpected rain when the king prays, protecting soldiers in their bronze armor from destruction.

After these events, as the years pass, this great and righteous king's rule will come to an end. Before he dies, he will name his golden-haired son as his successor. This new ruler, whose name contains two tens, will inherit the kingdom. He will be a brilliant leader, quick-minded, and as strong as the legendary Hercules. He will be a master in battle, hunting, and horseback riding, but he will live a lonely and dangerous life.

During his rule, a terrifying sign will appear: a thick mist will cover Rome's plains so completely that people will be unable to see their neighbors. War and deep sorrow will follow. The king, blinded by his passion, will bring disgrace upon himself by marrying inappropriately and dishonoring his children.

Then, he will face a tragic end—hiding alone, trapped by fate, he will die in a bathhouse, betrayed and defeated by the war god, Ares.

This will be a warning that Rome's downfall is near. Because of its endless hunger for power, many will be killed within the halls of the city. Rome will finally pay for everything it has done, repaying the destruction it once caused in its many wars.

My heart is heavy, and I am filled with sorrow.

From the time Rome's first king gave laws to the people and the Word of God came to earth, until the nineteenth ruler finishes his reign, two hundred and forty-four years will pass. Then, during the rule of the twentieth king, he will be struck down by a sword, spilling blood in the streets of Rome. His name will contain the letter linked to the number eighty, and he will be an old man. Soon, Rome will become a widow, losing its ruler.

During this time, many warriors will rise, bringing battles, bloodshed, and endless suffering. Armies will clash, men and horses will fall in the fields, and war will consume the land.

Another ruler will take power, with the number ten in his name. He will bring pain and hardship, robbing people of their wealth. But his rule will be short-lived, as he will fall in battle, killed by a mighty warrior's sword.

Then, a ruler connected to the number fifty will come from the East. A fierce warrior, he will go to Thrace, only to be driven away. He will flee to Bithynia and the plains of Cilicia, but war will catch up to him, and he will be defeated in Assyria.

After him, a deceitful leader will take control, skilled in trickery and schemes. He will rise from the West, and his name will have the number two hundred. He will start a war for power over the Assyrians, building a massive army and seizing control. He will rule the Romans with force, filled with ruthless ambition. A violent and greedy leader, he will destroy noble families, steal their wealth, and leave the land in ruins. His hunger for power will lead him eastward, spreading deception and chaos wherever he goes.

Then, a young ruler will share power, bearing the name of a great Macedonian leader. He will find himself caught in betrayal and deception, barely escaping a plot against him within his own army. However, the barbaric ruler who seeks to overpower him will meet a sudden death, struck down by a warrior's sword. Even after his death, the people will tear his body apart.

Soon, the kings of Persia will rise, and Roman warriors will battle once again.

Phrygia will suffer from earthquakes, and destruction will come upon Laodicea and Hierapolis. The earth will open, swallowing parts of the cities just as it did once before. The people will mourn as war spreads, and the world will be filled with suffering.

A ruler from the East will march toward Italy, but he will meet his end, falling to the sword. His downfall will be tied to his mother, and he will be remembered with hatred.

The seasons will be unpredictable, shifting in ways people do not understand. Some will suffer while others prosper. But those who honor God and turn away from idolatry will find peace.

Now, Lord of all things, the eternal King, you have placed this prophecy in my heart. I ask you to let me rest now, for I do not fully understand the words I speak. Let this vision end, for my heart is weary from revealing the future of kings and their rise and fall.

Book 13.

The eternal God, who never changes, has commanded me to speak again. He is the one who gives power to kings and takes it away. He decides the length of their rule and when their lives will end.

Even though I do not wish to deliver this message, I must warn kings about what is to come.

War and destruction will spread, led by violent warriors. Many will die—both children and elders who guide their people with wisdom. Battles will rage, bringing hunger, disease, earthquakes, and powerful storms. Armies from Assyria will roam across the land, looting cities and stealing from temples.

A rebellion will rise among the hardworking Persians, joined by the Indians, Armenians, and Arabians. In response, a Roman king, hungry for war, will lead his soldiers into battle against the Assyrians. A young warrior, skilled in combat, will push forward as far as the Euphrates River. However, he will be betrayed by a trusted friend and fall in battle, struck down by a soldier's sword.

Then, a ruler from Syria, who loves wealth and power, will take control. He will be ruthless in war, and his son will follow in his footsteps, ruling the world with a heavy hand. Both will share the same name, and their combined rule will be marked by the numbers five hundred, one, and twenty. While they are in power, laws will be passed, and there will be a short time of peace. But it will not last.

A deceiver will rise, pretending to make peace with his enemies. Like a wolf promising to protect a flock of sheep, he will take an oath only to break it. He will betray and destroy those who trusted him, throwing away his promises.

Then, war will break out again between proud and greedy kings. Many people will suffer, including the Syrians, Indians, Armenians, Arabians, Persians, and Babylonians. These nations will destroy each other in terrible battles.

When a powerful Roman warrior defeats a fierce German leader, war will continue for many years among the Persians. But they will not find victory. Just as a fish cannot climb a steep mountain, a turtle cannot fly, and an eagle cannot live in the sea, so too will the Persians struggle and fail in battle.

Meanwhile, the land of the Italians, resting near the sacred waters of the Nile, will send aid to the seven-hilled city of Rome. As long as Rome's name remains in the pages of history, the great city built by the Macedonian ruler will provide food to its people.

I will now tell of terrible suffering for the people of Alexandria, who will be destroyed because of the cruelty of wicked rulers. Once-powerful men will become weak and beg for peace as corruption spreads among their leaders.

God's anger will fall upon the Assyrians, and a great flood will wipe them out, reaching even the city of Caesar and harming the people of Canaan.

The Pyramus River will flood the city of Mopsus, and the people of Aegea will fall because of violent conflicts among strong warriors.

Antioch will not escape suffering, as a great war will surround it. A powerful leader will rise within its walls and battle against the Persian archers. He will gain control over Rome and lead its armies.

Cities of Arabia will be filled with grand temples, open markets, and great wealth. Statues of gold, silver, and ivory will be everywhere. Bostra and Philippopolis, known for their love of learning, will later suffer great sorrow.

Even the stars and signs in the sky will not help them. Aries, Taurus, Gemini, and the other constellations will not bring any relief. Those who put their trust in such things will find themselves betrayed.

Now, I will speak of Alexandria's wars. Its people will turn against one another, and many will die as their own city is destroyed. They will fight for power and greed, and the god of war will rage among them. A leader with a strong heart, along with his mighty son, will fall in betrayal because of an older ruler.

After this, another powerful ruler will take over fertile Rome. He will be skilled in war and come from the Dacians. His name will carry the number three hundred, along with the number four. He will kill many people, including his own brothers and close friends, while other kings are slaughtered. Chaos, robbery, and murder will follow because of the older ruler's death.

Then, a cunning man will appear, a thief and a stranger to Rome, coming from Syria. Using deception, he will invade Cappadocia, bringing destruction. Tyana and Mazaka will be conquered, and their people will be enslaved again, forced to carry heavy burdens. Syria will mourn the loss of its people, and even the goddess of the moon will not protect her city.

When this ruler flees from Syria and faces the Romans, he will no longer fight like them but will take on the ways of the Persian archers. Then, fulfilling fate, the ruler of Italy will fall in battle, struck down by a shining sword. Soon after, his children will perish as well.

A new ruler will take control of Rome, but soon, unstable nations will rise against the empire. Rome's walls will be surrounded by war and destruction. Famine, disease, powerful storms, and terrible battles will follow. Cities will fall into chaos, and many Syrians will be wiped out.

God's great wrath will strike them, and the Persians will join the Syrians in attacking the Romans. However, they will not be able to fully conquer the land or destroy its laws.

Many people will flee from the East, seeking safety in foreign lands. The ground will be soaked in the blood of countless victims. It will be a time when the living will envy the dead, and people will see death as a blessing—but death will not come for them.

Now, I mourn for Syria, for a terrible disaster will strike its people. Deadly arrows will rain down upon them from an enemy they never expected.

A fugitive from Rome will arrive, leading a vast army and carrying a great spear. Crossing the Euphrates River with thousands of soldiers, he will burn Syria to the ground and leave it in ruins.

Antioch, your fate is sealed. You will no longer be called a city, for your lack of wisdom will lead to your downfall. Your streets will be filled with enemy soldiers, and your homes will be stripped bare. Left with nothing, you will become an empty shell, abandoned and ruined.

Many will weep suddenly when they see the destruction.

Hierapolis, you will be taken in victory, and so will you, Beroea. Chalcis, you will cry for your sons who have been wounded in battle.

So many will suffer near the steep mountains of Casius and Amanus. So many will be lost near the rivers Lycus, Marsyas, and the silver-flowing Pyramus. The conquerors will take everything from Asia, stealing treasures, leaving cities in ruins, and destroying temples.

One day, sorrow will fall upon the Gauls, Pannonians, Mysians, and Bithynians when a mighty warrior arrives.

Lycians, beware—a fierce enemy, like a wolf, will come to spill your blood. The Sannians will invade, bringing destruction, while the Carpians and Ausonians prepare for war.

A king's illegitimate son will betray him, taking his life through treachery. But he, too, will soon die because of his own wickedness.

Another ruler will follow, his name beginning with the first letter, but he will not last long either, falling in battle by a warrior's sword.

The world will fall into chaos once again. People will die from both war and disease. The Persians, angered by the Ausonians, will return to battle, causing the Romans to flee.

A priest, famous everywhere, will rise from Syria, acting with deception to achieve his goals. The city of the sun will pray for protection, while the Persians will threaten the Phoenicians.

When two strong leaders take control of Rome, one with the number seventy and the other with three, chaos will follow. A mighty bull will charge, kicking up dust with its hooves, attacking a dark serpent that slithers on the ground. But in the end, the bull itself will fall.

After him, a swift and hungry stag will roam the mountains, searching for food among dangerous creatures.

Then, a fierce lion will come, sent by the sun, breathing fire. It will destroy the powerful stag and strike down a deadly serpent that makes a terrible hissing sound. The sideways-moving goat will fall as well, and this lion will gain great fame.

The lion will rule over Rome, and the Persians will grow weak.

Lord, ruler of the world, let this song come to an end and bring peace to our words.

Book 14.

O people, why do you dream of things too great, as if you will live forever?

You rule for only a short time, yet you desire power over everyone. You do not understand that God despises greed for power and hates wicked rulers who crave control. Because of this, he sends darkness upon them, and instead of doing what is right, they dress in royal robes and seek war and bloodshed. But God, who is eternal, will shorten their reigns and bring them to ruin, one after another.

Then a fierce leader will rise, strong and wild, destroying everything in his path. He will tear apart even the shepherds, and no one will stop him unless fast and clever young warriors chase him down through the forests. A great battle will come when a brave fighter hunts down the beast that has terrorized the people.

After this, another ruler will take power, a man with a name of four syllables, and his name will begin with the first number. But war will quickly bring him down.

Then two leaders, both connected to the number forty, will rule together. Under them, peace and justice will spread across the land. However, greedy men, hungry for gold and silver, will betray them and take their lives using cunning plans.

After them, a young and fierce ruler, marked by the number seventy, will rise. He will betray Rome's army and cause great suffering. Because of his actions, cities and homes will be destroyed, and Rome will fall. The once-great city will be nothing but ashes, with nothing left of its former glory.

Then, from the sky, God will send fire and destruction. He will strike the wicked with lightning and thunder, burning some and crushing others. The cruel ruler who caused Rome's downfall will be killed by the very people he betrayed. His body will not be buried with honor, but left for the birds and wild beasts to devour.

After him, another leader, known for conquering the Parthians and Germans, will take power. He will hunt and destroy the wild beasts that threaten people along the oceans and the Euphrates River. Under him, Rome will regain its strength.

But soon, a fierce enemy will come, like a wolf invading the land. He will march from the West, but before he can rule for long, he will be struck down by a warrior's sword.

Another great leader, from Assyria, will rise. His name will start with the first letter, and he will try to bring order through war. But treacherous forces will turn against him, and he will fall in battle.

After him, three powerful rulers will come. One will have the first number, another the number thirty, and the last will be connected to three hundred. They will be cruel men who melt gold and silver to create false idols. To win wars, they will use money to bribe armies, giving out treasures and riches.

They will fight against the Parthians, the Medes, and the strong warriors of Persia. When one of them dies, he will leave his kingdom to his sons, hoping they will rule wisely. But instead, they will ignore his words and fight among themselves for power.

Then another ruler, tied to the number three, will take control, but he will not last long before he is struck down by the sword. After him, many will fight and kill one another, each trying to claim the throne.

A strong and wise leader will rule over the powerful Romans. He will be an older man, connected to the number four, and will govern well.

Then, war will come to Phoenicia as Persian archers attack. Many people will fall to these invaders who speak foreign tongues. The cities of Sidon, Tripoli, and Berytus, once proud and strong, will see their streets covered in blood and bodies.

Laodicea will bring upon itself a terrible and hopeless war because of the sins of its people.

The people of Tyre will suffer greatly. In the middle of the day, the sun will disappear, and dark red rain will fall from the sky. During this time, the king will be betrayed by those closest to him. After his death, corrupt leaders will rise, continuing the cycle of violence and killing each other for power.

Then, a wise and respected ruler will take the throne. His name will be tied to the number five, and he will command a strong army. The people will admire him for his leadership, and he will build a good reputation through his actions. But during his reign, a terrible event will occur between the Taurus and Amanus mountains. A new, strong, and beautiful city in Cilicia, near powerful rivers, will be destroyed. Many earthquakes will also shake Propontis and Phrygia. In the end, this great king will die from a deadly illness.

After him, two kings will rule. One will be connected to the number three hundred, and the other to the number three. They will destroy many enemies to defend the seven-hilled city of Rome and maintain their control. But trouble will come to the senate, as the ruling king will hold deep anger against it. A strange sign will appear for all to see, and the world will experience heavy rain, snow, and hail, ruining crops. These kings will ultimately die in wars fought for Italy.

Then, another ruler will rise—a cunning man who will gather a great army. To prepare for war, he will give money to soldiers wearing bronze armor. But during his reign, the Nile River will flood beyond the Libyan lands, bringing water to Egypt for two years. Despite this, famine, war, and crime will spread, leading to the destruction of many cities. In the end, this ruler will be betrayed and killed by the sword.

After him, a leader tied to the number three hundred will take control of Rome and its mighty warriors. He will lead brutal attacks against the Armenians, Parthians, Assyrians, and Persians. During his reign, Rome will be rebuilt with gold, silver, ivory, and amber, making it grander than ever. People from both the East and the West will gather there, and the ruler will create new laws. But in the end, he will meet his fate on a distant island.

Then, another ruler will take power, connected to the number thirty. He will be fierce, wild-haired, and descended from the Greeks. His rule will bring destruction to the cities of Molossian Phthia and Larissa along the Peneus River.

At the same time, an uprising will break out in Scythia, and war will rage near Lake Maeotis and the river Phasis. Many warriors will die in battle. The king will defeat the Scythians, but after his victory, his life will come to an end.

Another ruler, connected to the number four, will take power. He will be a terrifying leader, feared in battle by the Armenians, who drink the cold waters of the Araxes River, and by the strong-willed Persians. War will break out between the Colchians and the mighty Pelasgians. The people of Phrygia and those near the Propontis will also fight, drawing their swords in violent battles driven by wickedness.

Then, with time, God will send a great sign from the sky—a bat, warning of an upcoming war. The king will not escape his fate; he will be struck down by a sword and die.

After him, another ruler, tied to the number fifty, will rise from Asia. He will be a fierce warrior, engaging in hand-to-hand combat. He will bring war to the grand walls of Rome and battle the Colchians, Heniochi, and the nomadic Agathyrsians by the Black Sea and the shores of Thrace. But his end will be brutal—he will be slain, and his body will be torn apart.

With the death of the king, Rome, once great and filled with people, will become a desolate land. Many will perish.

Then, a terrifying ruler will emerge from Egypt. He will defeat the powerful Parthians, Medes, and Germans, along with the Agathyrsians of the Bosporus, the Iernians, the Britons, and the Iberians. He will also fight the Massagetæ, skilled archers, and the Persians, who think themselves invincible.

A renowned leader will rise against all of Greece, treating Scythia and the towering Caucasus Mountains as his enemies.

During his reign, a strange sign will appear—crowns that shine like stars will be seen in the sky over both the northern and southern lands. When he dies, he will pass his power to his son, whose name starts with the first letter of the alphabet. The king will meet his fate and descend into the underworld.

When this son takes control of Rome, his rule—marked by the number one—will bring a long-awaited peace across the world. The Latins will cherish him as their king, remembering the legacy of his father. Though he will long to travel east and west, the Roman people will hold him back, insisting he remain as their ruler. They will feel a deep loyalty to their noble leader.

However, death will come swiftly, taking him too soon. Afterward, chaos will return as powerful warriors fight among themselves, no longer ruling as kings but as cruel tyrants. Across the world, they will bring suffering, especially to the Romans, until the arrival of a third leader named Dionysus.

Then, from Egypt, a warlike ruler will rise, known as Dionysus the Lord. When the royal purple cloak is torn apart by a violent lion and lioness, the kingdom will shift. But a righteous king, whose name begins with the first letter of the alphabet, will rise up. He will crush his enemies, leaving their bodies to be devoured by dogs and birds.

Oh, Rome, once powerful, you will be burned by fire. So much suffering awaits you when these events take place. But later, a great and famous ruler will rebuild you, using gold, silver, amber, and ivory. Once again, you will be rich in temples, marketplaces, and arenas. You will shine like before, a guiding light for all.

But woe to the lands of Cecropes, Cadmeans, and the Spartans—those who live near the Peneus River and the Molossian waters, the cities of Tricca, Dodona, Ithome, and the peaks of Olympus, Ossa, Larissa, and Calydon.

A great sign will appear—a darkened day like twilight covering the earth. This will mark the end for a powerful king, who will be struck down by his own brother's arrow. Afterward, another ruler will rise, a fiery leader from a royal bloodline. He will seize power over Egypt's people. Though younger than his brother, he will be much stronger. His name will be linked to the number eighty.

Then, the whole world will face the unstoppable anger of God. Humanity will suffer from famine, plagues, war, and endless violence. Darkness will cover the earth, bringing destruction. From the sky, storms, hail, and fiery thunderbolts will strike, while the ground shakes beneath Scythian hills and Greek cities. Many places will crumble under God's wrath, set ablaze by burning lightning.

The great ruler will not escape fate—his own men will kill him as if he were nothing. After him, many men from the Latin people will wear the purple robe of power. They will fight for control, each one eager to rule.

Three kings will rise in Rome, two connected to the first number, and one with a name meaning victory. They will love Rome and the world, hoping to bring peace to its people. But they will fail because God will not be merciful. Humanity has done too much evil, and punishment will continue.

God will allow cruel leaders to take control—men even more vicious than wolves or leopards. They will be ruthless, betraying and killing kings like helpless victims. Rome's great leaders, believing in false promises, will be destroyed.

Warriors, chaotic and full of rage, will attack without order. They will spill the blood of noble families and first-born sons, bringing Rome to ruin. Three times, the Most High will bring terrible judgment upon the world, destroying both people and their wicked works. But those with shameless hearts, who have committed evil, will still be judged. They will be trapped, falling upon one another, and condemned for their wickedness.

A bright comet will signal what is to come—war, destruction, and great battles.

During these troubled times, a leader will collect prophecies from distant lands, predicting disaster for temples and cities. He will order Rome to store up wheat and barley for twelve months, preparing for hardship. The city will suffer greatly, but then it will recover. Peace will return only when this ruler is gone.

After that, the final race of Latin kings will rule. But after them, a new kingdom will rise, strong and unshaken. It will be clear to all that God himself is the true king.

There is a special land, rich and fertile, lying in a great plain. The Nile River surrounds it, separating it from Libya and Ethiopia.

The Syrians, scattered from different places, will take everything they can carry. A strong and wise ruler will take charge, training young men and sending soldiers to battle. He will focus on a terrifying war and send a powerful ally to help all of Italy. When he reaches the dark sea near Assyria, he will attack the Phoenicians in their homes, bringing destruction and fierce battles. He will rule as one of the two most powerful leaders on earth.

Now I must speak of the terrible fate awaiting Alexandria. Foreign invaders will take over Egypt, a land that once stood strong and untouched. When the gods' anger falls upon it, there will be great suffering.

A great change will come, making winter feel like summer. Prophecies will be fulfilled. When three young men win at the Olympian games, they will be asked to perform a cleansing ritual using the blood of a newborn animal. But this will not stop what is coming. Three times, the Most High will bring disaster. A heavy spear will be raised over all people, and blood will flood the ground when a city is completely destroyed by cruel invaders.

Blessed are those who are already dead, and even more blessed are those without children. The leader, once known for his freedom, will abandon his past ideals and force his people into slavery. A new ruler will bring great sorrow. Soon, a doomed army of Sicilians will arrive, bringing even more destruction. Another foreign army will strike, cutting down crops before they can be harvested. The great thunderous God will punish them, and people will fight over stolen gold, taking it from one another.

When people witness the fall of the powerful lion, a deadly lioness will follow, bringing even more destruction. The ruler's power will be stripped away. Just as people in Egypt celebrate feasts and cheer loudly, so too will the world be filled with the noise of war, terror, and destruction. Many will die, and people will turn against each other in brutal battles.

Then a figure covered in dark scales will rise, joined by two others working together. A third leader, a great warrior from Cyrene, once a refugee in Egypt, will also rise. However, none of their plans will succeed.

For a time, the world will be peaceful, but another war will soon break out in Egypt. A great naval battle will take place, but the victors will not hold onto their power for long. A famous city will be conquered, but only briefly.

People from neighboring lands will flee, leading their families to safety. But when they return, they will fight for victory. The Jews, strong warriors, will battle fiercely to protect their homeland and loved ones. Great warriors will be counted among the dead. Many will drown in the sea, and the shores will be covered in bodies. Birds will feast on golden-haired soldiers who have fallen in battle.

Arabian lands will be soaked in blood. When wolves and dogs swear false oaths on an island, a great tower will be built. A once-ruined city will be inhabited again. Gold and silver will lose their value, and people will no longer fight over wealth. Servitude will end. People will live together in friendship, sharing everything equally. Evil will vanish from the world, sinking deep into the sea.

The time of judgment for humanity will arrive. These events must happen. But even in those days, no traveler will say that humanity is coming to an end. A holy nation will rise and rule the world forever, with strong and noble leaders guiding them.

Thank You for Reading

Dear Reader,

We hope this timeless classic has sparked your imagination and enriched your literary journey. Now that you've turned the final page, we want to share a vision for the future of reading—one where every classic you've ever wanted to explore is at your fingertips, in a format that best suits your life.

We'd like to invite you to gain immediate, unlimited digital & audiobook access to hundreds of the most treasured literary classics ever written—along with the option to secure deluxe paperback, hardcover & box set editions at printing cost. Together, we can spark a new global literary renaissance alongside our small, independent publishing house called "The Library of Alexandria."

Thousands of years ago, the Library of Alexandria stood as a beacon of knowledge—until it was lost to history. We aim to reignite that spirit of preservation and discovery right now, in the modern age—only this time, it's accessible to all, in every language and every format.

Picture a world where every timeless classic, novel, poem, or philosophical treatise is not only available to read but also updated for today's readers—modernized, translated into any language or dialect, and ready to enjoy in any format you choose, whether that is in an eBook, audiobook, paperback, or deluxe hardcover & box set version a printing cost.

By joining our movement to rebuild the modern Library of Alexandria, you become part of an unprecedented mission to offer:

- **Unlimited Audiobook & eBook Access to the Greatest Classics of All Time**

  Instantly explore thousands of legendary works, from Plato and Shakespeare to Jane Austen and Leo Tolstoy. All are instantly ready to read or listen to, giving you a complete literary universe at your fingertips.

- **Paperback & Deluxe Editions at Printing Costs:**

  Purchase any title in a paperback, deluxe hardbound, or deluxe boxset edition at printing costs, shipped right to your doorstep. Curate your personal library of Alexandria with editions worthy of display—crafted to last, designed to captivate, and delivered straight to your door.

- **Modern translations for Contemporary Readers in all languages and dialects**

  Discover a vast selection of classics reimagined in clear, current language—no more struggling with outdated phrases or obscure references. Next to the original versions, we aim to offer translations in as many languages and dialects as possible.

  As we continue our translation efforts and add new languages, readers everywhere can connect with these works as if they were written today. By bridging linguistic divides, you're contributing to ensuring that these timeless stories become more meaningful, accessible, and inspiring for people across the globe.

- **Your Personal Library of Alexandria:**

  Over the months and years, you'll curate a unique physical archive of classics—each volume a testament to your taste, curiosity, and love of knowledge. It's not just about owning books—it's about curating a cultural legacy you'll cherish and pass down for generations to come.

- **Join a Global Literary Renaissance:**

  Your support fuels an ongoing mission: allowing us to reinvest in offering deluxe print editions (including special boxsets) at their true cost, broaden the range of available formats and translations, and extend the reach of these works to new audiences worldwide. By joining today, you're not just preserving a legacy of masterpieces; you set in motion a powerful wave of literary accessibility.

  We are more than a publisher—we're a movement, and we can't do it alone. Your support lets us scale our mission, preserving and reimagining history's greatest works for tomorrow's readers.

Become a Torchbearer of knowledge.

Thank you for picking up this book and allowing us into your literary journey. As you turn the pages, know that you're part of something larger: a global effort to keep these stories alive, share their wisdom across borders and generations, and spark a true cultural revival for the modern era.

If this resonates with you—please consider taking the next step by visiting:

www.libraryofalexandria.com

With gratitude and a shared love of knowledge,

The Modern Library of Alexandria Team

Visit:

www.libraryofalexandria.com

Or scan the code below:

www.ingramcontent.com/pod-product-compliance
Lightning Source LLC
La Vergne TN
LVHW040542210926
848483LV00003B/396
* 9 7 8 1 8 0 4 2 1 7 9 4 8 *